I0485839

# Industrial Chemistry

:: Author ::

Dr. Darshan V. Chaudhary

## PUBLISHED BY

The New Era International Publishing House
H.Q. At & Po. Chaveli., Ta- Chansma,
Dist- Patan, North Gujarat, India, Asia.
www.iphouseindia.com

First Publication: 5th September, 2015

Copyright: Author

(c) **Dr. Darshan V. Chaudhary**

ISBN:- 978-1-51722-015-0

Price: Rs.800/- INDIA

   $ 10 OUTSIDE INDIA

**PUBLISHED BY**

**The New Era International Publishing House**
**H.Q. At & Po. Chaveli., Ta- Chansma,**
**Dist- Patan, North Gujarat, India, Asia.**
**www.iphouseindia.com**

# PREFACE

Industrial chemistry contracts with marketable manufacture of chemicals and connected products from natural rare resources and their derivatives. It permits mortality to involvement the benefits of chemistry when we put on it on the manipulation of resources and energy. When we apply chemistry in the conversion of materials and energy to brand practical products, this outcomes in development and perfection in areas such as food production, healthiness and cleanliness, shelter and clothing. The financial growth of industrialized countries trusts on the manufacturing industry for complete products. The aim of learning industrial chemistry at academic is to attempt and connection the break among established chemistry. The chemical industry is extremely globalized and manufactures thousands of chemicals from a extensive diversity of raw materials by incomes of diverse technologies for diverse end usages. It is consequently imperative to base the study of industrial chemistry on an empathetic of the structure of the industry and the unit operations and unit processes that make up the chemical processes.

In present book, chapter 1 introduces the scope of chemical industry, its location in the universal economy, and its arrangements in terms of the chemical procedures that describe it. To allow the study of nominated biochemical processes. Chapter 2 describes the unit operations and unit processes that are relavent in knowledge activities. Chapter 3 elobarates the study of industrial inorganic and organic chemical industries, size reduction and separation unit operations its also includes the study of extractive metallurgy of iron, copper and aluminium. Chapter 4 emphasis on some basic inorganic industrial developments that synthesize products from a variation of rare resources derived from the normal atmosphere. They contain production of chlorine and sodium hydroxide from brine, ammonia from methane and nitrogen, sulphuric acid from sulphur, fertilizer and cement from mineral ores. Chapter 5 discusses the study of organic industrial chemistry then starts with petroleum refining followed by the manufacture of selected petrochemicals and polymers. Chapter 6 elaborates the study of ethanol, pharmaceuticals, soaps and detergents.

The contents of the book will be useful to the students of Industrial Chemists, Biochemistry, Biotechnology, Pharmaceutical science and technology, Food science and technology, Organic Chemists etc.

I express my heartfelt thanks to Dr.Pranav S. Shrivastav, Professor, Chemistry Department, Gujarat University, Ahmedabad, India, for his constant guidance across my research Work and without the same platform I would not be able to compile this book. I am very thankful to my other colleague contributor Mr.EdvinPithawala for critical evaluation of each chapter across the book. I would like to express my gratitude to my family members especially to my parents for their love affection and care and last but not the least to my beloved *Reema* for her everlasting love, motivation and sacrifice for the time taken in compiling this book.

I am grateful to publisher for their concern, efforts and encouragement, especially for their excellent cooperation in the task of preparing and publishing this book.

- **Dr.DarshanChaudhary**

# TABLE OF CONTENTS

# CHAPTER – 1

## Introduction to Industrial Chemistry

☞ **CONTENTS**

## 1.7 BRIEF OUTLINE ABOUT THE WORK

## 1.8 SUMMARY

## 1.9 REFERENCES

### 1.1 The difference between classical and industrial chemistry

Before we define industrial chemistry, it may be helpful to know that the development of industrial chemistry started when a need to know how various chemicals are produced in much more than the laboratory scale, arose. Chemistry knowledge was applied to furnish the rapidly expanding chemical industries with ''recipes'' which we now call **chemical processes**. Industrial chemistry keeps up with the progress in science and technology. It incorporates other emerging disciplines such as biotechnology, microelectronics, pharmacology and material science. The discipline is also concerned with economics and the need to protect the environment.

We define industrial chemistry **as the branch of chemistry which applies physical and chemical procedures towards the transformation of natural raw materials and their derivatives to products that are of benefit to humanity.**

Classical chemistry (organic, inorganic and physical chemistry) is very essential for advancing the science of chemistry by discovering and reporting new products, routes and techniques. On the other hand industrial chemistry helps us to close the gap between classical chemistry as it is taught in colleges and universities, and chemistry as it is practiced commercially. The scope of industrial chemistry therefore includes:

☞The exploitation of materials and energy in appropriate scale

☞Application of science and technology to enable humanity experience the benefits of chemistry in areas such as **food production**, **health and hygiene,shelter, protection, decoration, recreation and entertainment.**

## 1.2 Classification of Industries

Industry is a general term that refers to all economic activities that deal with production of goods and services. Goods and services are key words when you talk of industry. We then expect industry to include the following **sectors:**

☞**Manufacturing**

Agriculture, Energy.Transport, Communication, Education,Tourism, Building and construction Trade, Finance etc

## 1.2.1 Classification of the Manufacturing Industry

The manufacturing industry is the area of focus in the study of this module. Manu-facturing produces manufactured goods. This makes it distinct from other sectors like agriculture which also produce goods. In manufacturing, materials are transformed into other more valuable materials.

We define manufacturing industry as follows:

**Manufacturing industry is a compartment of industry or economy which is concerned with the production or making of goods out of raw materials by means of a system of organized**

**labour.**

Manufacturing industry can be classified into two major categories namely, **heavyand light industry**.

• Capital-intensive industries are classified as heavy while labour intensive industries are classified as light industries.

• Light industries are easier to relocate than heavy industries and require less capital investment to build.

Using the above classification criteria, examples of heavy industries include those that produce industrial machinery, vehicles and basic chemicals.

Other measures used to classify industries include the weight or volume of products handled and weight per cost of production. For example the weight of steel produced per dollar is more than the weight per dollar of a drug. In this case, steel industry is a heavy industry whereas drug manufacture is a light industry.

Sometimes governments define heavy industry in terms of its impact on the environment. Many pollution control laws target heavy industries which in most cases pollute more than light industries. Therefore, pulp and paper industry is a heavy industry since its contribution to pollution is enormous.

Both inorganic and organic chemical industry can be either heavy or light industry. For example the pharmaceutical industry which is basically organic is light industry.

Petroleum refining is organic but heavy industry. Iron and steel industry is inorganic and heavy industry.

## 1.2.2 Manufacturing sub-sectors

Because the raw materials and the actual products manufactured are so varied, different skills and technologies are needed in manufacturing. Manufacturing is therefore divided into sub-sectors which typically deal with category of goods such as the following:

• Food, beverages and tobacco

• Textiles, wearing apparel, leather goods

• Paper products, printing and publishing

• **Chemical, petroleum, rubber and plastic products**

• Non-metallic mineral products other than petroleum products

   Basic metal products, machines and equipment.

Let us now focus on the **chemical, petroleum, rubber and plastic products sub-sector.** We shall generally call it the chemical industry.

## 1.3    The Chemical Industry

The chemical industry can also be classified according to the type of main raw ma-terials used and/or type of principal products made. We therefore have **industrialinorganic chemical industries** and **industrial organic chemical industries.** Industrial inorganic chemical Industries extract inorganic chemical substances, make composites of the same and also synthesize inorganic chemicals.

Heavy industrial organic chemical industries produce petroleum fuels, polymers, petrochemicals and other synthetic materials, mostly from petroleum.Light organic industries produce specialty chemicals which include pharmaceuticals, dyes, pigments and paints, pesticides, soaps and detergents, cosmetic products and miscellaneous products.

## 1.3.1 The Structure of the Global Chemical Industry

We normally put a value to something according to how much it has cost us. Some things are of high value while others are of low value. For low valued products, you need to produce them in large volumes to make significant profit. This means that the raw materials are cheap and easily accessible. There is also an existing, relatively simple, and easily accessible processing technology. To sell a large volume of prod-uct, there must be a large market. This brings stiff competition which also makes the price to remain low.

We are now ready to describe the structure of the global chemical industry

☞ **Commodity Chemicals**

The global chemical industry is founded on basic inorganic chemicals (BIC) and basic organic chemicals (BOC) and their intermediates. Because they are produced directly from natural resources or immediate derivatives of natural resources, they are produced in large quantities.

In the top ten BIC, almost all the time, sulphuric acid, nitrogen, oxygen, ammonia, lime, sodium hydroxide, phosphoric acid and chlorine dominate. The reason sulph-uric acid is always number one is because it is used in the manufacture of fertilizers, polymers, drugs, paints, detergents and paper. It is also used in petroleum refining, metallurgy and in many other processes. The top ranking of oxygen is to do with its use in the steel industry.

Ethylene and propylene are usually among the top ten BOC. They are used in the production of many organic chemicals including polymers.

BIC and BOC are referred to as commodity or industrial chemicals.

Commodity chemicals are therefore **defined** as low-valued products produced in large quantities mostly in continuous processes. They are of technical or general purpose grade.

☞ **Specialty Chemicals**

High-value adding involves the production of small quantities of chemical products for specific end uses. Such products are called specialty chemicals. These are high value-added products produced in low volumes and sold on the basis of a specific function.

In this category are the so-called **performance chemicals** which are high value products produced in low volumes and used in extremely low quantities. They are judged by performance and efficiency. Enzymes and dyes are performance chemicals.

Other examples of specialty chemicals include **medicinal chemicals, agrochemi-cals, pigments, flavour and fragrances, personal care products, surfactants and adhesives.**

Specialty chemicals are mainly used in the form of formulations. Purity is of vital importance in their formulation. This calls for organic synthesis of highly valued pure chemicals known as **fine chemicals**.

☞ **Fine Chemicals**

At times you will find that the raw materials for your product need to be very pure for the product to function as desired. Research chemicals are in this category as also are pharmaceutical ingredients. Such purified or refined chemicals are called fine chemicals. By **definition** they are high value-added pure organic chemical substances produced in relatively low volumes and sold on the basis of exact specifications of purity rather than functional characteristics.

The global market share for each type is roughly as follows:

**Commodities    80%**

Specialties    18%

**Fine            2%**

### 1.4    Raw material for the Chemical Industry

We have paid some attention to products from the chemical industry. But, since there would be no chemical industry without raw materials, the subject of raw materials is due for discussion at this stage. All chemicals are derived from raw materials available in nature. The price of chemicals depends on the availability of their raw materials. Major chemical industries have therefore developed around the most plentiful raw materials

The natural environment is the source of raw materials for the chemical industry.

### Raw materials from the atmosphere

The atmosphere is the field above ground level. It is the source of air from which six industrial gases namely $N_2$, $O_2$, Ne, Ar, Kr and Xe are manufactured. The mass of the earth's atmosphere is approximately $5x \ 10^{15}$ tons and therefore the supply of the gases is virtually unlimited.

## Raw materials from the hydrosphere

Ocean water which amounts to about $1.5 \times 10^{21}$ litres contains about 3.5 percent by mass dissolved material. Seawater is a good source of sodium chloride, magnesium and bromine.

## Raw materials from the lithosphere

The vast majority of elements are obtained from the earth's crust in the form of mineral ores, carbon and hydrocarbons. Coal, natural gas and crude petroleum besides being energy sources are also converted to thousands of chemicals.

## Raw materials from the biosphere

Vegetation and animals contribute raw materials to the so-called agro-based industries. Oils, fats, waxes, resins, sugar, natural fibres and leather are examples of thousands of natural products.

## Chemical Processes

Every industrial process is designed to produce a desired product from a variety of starting raw materials using energy through a succession of treatment steps integrated in a rational fashion. The treatments steps are either physical or chemical in nature.

*A Typical Industrial Process*

Energy is an input to or output in chemical processes.

The layout of a chemical process indicates areas where:

- raw materials are pre-treated

- conversion takes place

- separation of products from by-products is carried out

- refining/purification of products takes place

- entry and exit points of services such as cooling water and steam

## 1.4.1. Units that make up a chemical process

A chemical process consists of a combination of chemical reactions such as synthesis, calcination, ion exchange, electrolysis, oxidation, hydration and operations based on physical phenomena such as evaporation, crystallization, distillation and extraction. A chemical process is therefore any single processing unit or a combination of processing units used for the conversion of raw materials through any combination of chemical and physical treatment changes into finished products.

### 1.4.1.1. Unit processes

Unit processes are the chemical transformations or conversions that are performed in a process. In Table 1.1, examples of some unit processes are given.

**Table 1.1** Examples of unit processes

| Acylation | Calcinations | Dehydrogenation | Hydrolysis |
|-----------|--------------|-----------------|------------|
| Alcoholysis | Carboxylation | Decomposition | Ion Exchange |
| Alkylation | Causitization | Electrolysis | Isomerization |
| Amination | Combustion | Esterification | Neutralization |
| Ammonolysis | Condensation | Fermentation | Oxidation |

| | | | |
|---|---|---|---|
| Aromatization | Dehydration | Hydrogenation | Pyrolysis |

## 1.4.1.2. Unit Operations

There are many types of chemical processes that make up the global chemical industry. However, each may be broken down into a series of steps called **unit operations.** These are the physical treatment steps, which are required to:

•    put the raw materials in a form in which they can be reacted chemically

•    put the product in a form which is suitable for the market In Table1.2, some common unit operations are given.

**Table 1.2** Examples of unit operations

| | | |
|---|---|---|
| Agitation | Dispersion | Heat transfer |
| Atomization | Distillation | Humidification |
| Centrifuging | Evaporation | Mixing |
| Classification | Filtration | Pumping |
| Crushing | Flotation | Settling |
| Decanting | Gas absorption | Size reduction |

It is the arrangement or sequencing of various unit operations coupled with unit processes and together with material inputs, which give each process its individual character. The individual operations have common techniques and are based on the same scientific principles. For example, in many processes, solids and fluids must be moved; heat or other forms of energy may be transferred from one substance to another; drying, size reduction, distillation and evaporation are performed. By studying systematically these **unit operations**, which cut across industry and process lines, the treatment of all processes is unified and simplified.

## 1.4   Flow Diagrams

*A picture says more than a thousand words*

Some chemical processes are quite simple; others such as oil refineries and petroche-mical plants can be very complex. The process description of some processes could take a lot of text and time to read and still not yield 100% comprehension. Errors resulting from misunderstanding processes can be extremely costly.

To simplify process description, flow diagrams also known as flow sheets are used. **Aflow diagram is a road map of the process, which gives a great deal of information in a small space.** Chemical engineers use it to show the sequence of equipment andunit operations in the overall process to simplify the visualization of the manufacturing procedures and to indicate the quantities of material and energy transferred.

A flow diagram is not a scale drawing but it:

•      pictorially identifies the chemical process steps in their proper/logical sequence

•      includes sufficient details in order that a proper mechanical interpretation may be made

Two types of flow diagrams are in common use, namely, the block diagrams and the process flow diagrams.

### 1.5.1. Block Diagrams

This is a schematic diagram, which shows:

•    what is to be done rather than how it is to be done. Details of unit operations/processes are not given

•    flow by means of lines and arrows

•    unit operations and processes by figures such as rectangles and circles raw materials, intermediate and final products

### 1.5.2. Process flow diagram / flow sheet

Chemical plants are built from process flow drawings or flow sheets drawn by chemical engineers to communicate concepts and designs. Communication is impaired if the reader is not given clear and unmistakable picture of the design. Time is also wasted as reader questions or puzzles out the flow diagram. The reader may make serious mistakes based on erroneous interpretation of the flow diagram.

Communication is improved if accepted symbols are used. The advantages of correct use of symbols include:

•    the function being performed is emphasized by eliminating distractions caused by detail

•    possibility of error that is likely to occur when detail is repeated many times is virtually done away with

•    equipment symbols should neither dominate the drawing nor be too small for clear understanding.

Flow sheet symbols are pictorial quick-to-draw, easy-to-understand symbols that transcend language barriers.

Some have already been accepted as national standards while others are symbols commonly used in chemical process industries, which have been proven to be effective. Engineers are constantly devising their own symbols where standards do not exist. Therefore, symbols and presentation may vary from one designer or company to another.

Below is a cement process flow diagram illustrating the use of equipment symbols.

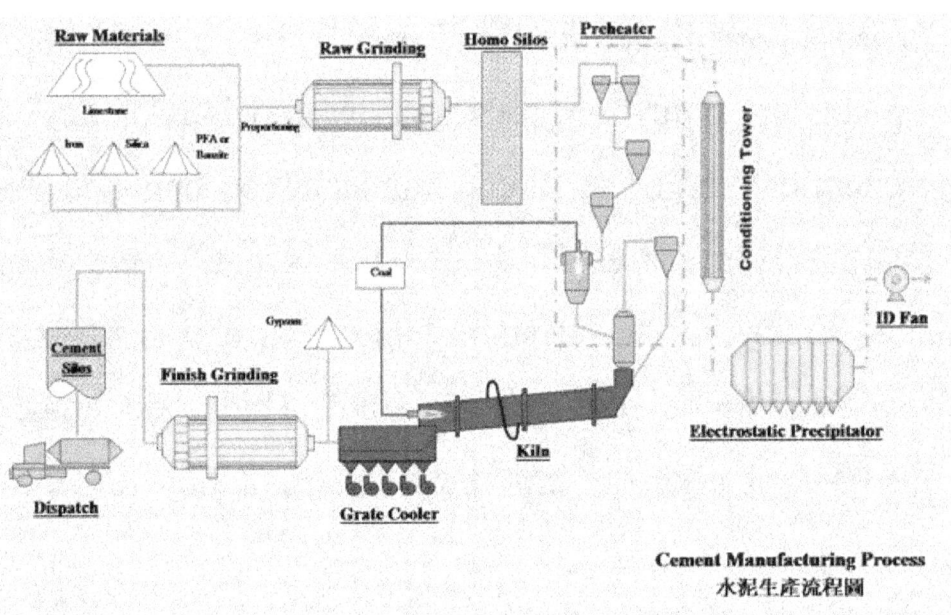

**Fig 1.1**   A process flow diagram for the manufacture of cement.

## 1.6   Material Balances

**From the law of conservation of mass, we know that m**ass can neither be creatednor destroyed. However, in nuclear reactions, mass and energy can be converted into each other respectively. Because of this, we can write equations called **"mass balances"** or **"material balances"**. Any process being studied must satisfy balances on the total amount of material, on each chemical component, and on individual atomic species.

As we have seen in the study of process diagrams, a process can have few or many streams depending on its complexity.

### 1.6.1 The purpose of mass balance calculations

Mass balance calculations serve the following purposes:

1.  They help us know the amount and composition of each stream in the pro-cess.

2.  The calculations obtained in 1 form the basis for energy balances through the application of the **law of conservation of energy.**

3.  We are able to make technical and economic evaluation of the process and process units from the knowledge of material and energy consumption and product yield obtained.

4.  We can quantitatively know the environmental emissions of the process.

In mass balance calculations, we begin with two assumptions

*   There is no transfer of mass to energy

*   Mass is conserved for each element or compound on either molar or weight basis

**It is important to note the following:**

*   Mass and atoms are conserved

*   Moles are conserved only when there is no reaction

*   Volume is not conserved.

You may write balances on total mass, total moles, mass of a compound, moles of an atomic species, moles of a compound, mass of a species, etc.

### 1.6.2 Material Balance Equations

We might be tempted to think that in a process,

**INPUT = OUTPUT**

In practice, some material may accumulate in the process or in some particular pro-cess units. For example, in a batch process, some material may remain adhered to the walls of containers. In the dehydration of ethane to ethylene, possible chemical reactions are as follows:

$$C_2H_{6(g)} \longrightarrow C_2H_{4(g)}$$
$$C_2H_{6(g)} \longrightarrow 2C_{(s)} +3H_{2(g)}$$
$$C_2H_{4(g)} \longrightarrow 2C_{(s)} +2H_{2(g)}$$

The carbon formed accumulates in the reactor.

Because processes may be batch with no inflow and outflow or continuous with inflow and outflow, and that there may be conversion of chemical species, a good mass balance equation takes care of all these aspects. The following is a general mass balance equation.

**Accumulation within the system**

**= Flow In through the system boundaries**

**- Flow Out through the system boundaries**

**+    generation within the system**

**- Consumption within the system**

Simply put:

**Accumulation =Flow in – Flow out + Production – Consumption**

The *system* is any process or portion of a process chosen for analysis. A system is said to be "open" if material flows across the system boundary during the interval of time being studied; "closed" if there are no flows in or out.

*Accumulation* is usually the rate of change of holdup of material within the system. If material is increasing, accumulation is positive; if it is decreasing, it is negative. If the system does not change with time, it is said to be at *steady state*, and the net accumulation will be zero.

The generation and consumption of material are the consequences of chemical reactions. If there is no chemical reaction, the production and consumption terms are typically zero.

**1.6.3 Mass balance calculation procedure**

The general procedure for carrying out mass balance calculations is as follows:

1. Make a block diagram (flow sheet) over the process
2. Put numbers on all the streams
3. List down all the components that participate in the process.
4. Find the components that are in each stream and list them adjacent to the stream in the block diagram
5. Decide on an appropriate basis for the calculations e.g. 100kg raw material A, 100kg/hr A, 1 ton of product, 100 moles reactant B etc.

6.  Find out the total number of independent relations. This is equivalent to the total number of stream components.

7.  Put up different relations between stream components and independent relations to calculate concentrations

8.  Tabulate results.

**Formative Evaluation**

1.  Distinguish between industrial and classical chemistry

2.  What factors are used to classify an industry as heavy or light?

3.  Define specialty chemicals

4.  Explain how the lithosphere is an important source of natural raw materials for the chemical industry

5.  What is the difference between unit operations and unit processes?

6.  What information would you expect to find in a block diagram for a chemical process?

7.  Discuss the use of symbols in process flow diagrams

8.  What assumptions are made at the initial stages of carrying out material ba-lance for a chemical process?

9.  Write the general mass balance equation

10. Producer gas has the following composition by volume:

|       | %    | Density, kg/m$^3$ |
|-------|------|-------------------|
| CO    | 28.0 | 1.2501            |
| CO$_2$ | 3.5  | 1.9768            |
| O$_2$  | 0.5  | 1.4289            |
| N$_2$  | 68.0 | 1.2507            |

The gas is burned with oxygen according to the following equation:

$$2CO + O_2 \longrightarrow 2CO_2$$

The oxygen is from the air whose volumetric composition is assumed to be 80% $N_2$ and 20% $O_2$. The oxygen fed from the air and the producer gas is 20% in excess of the amount required for complete combustion. The combustion is 98% complete.

Carry out total material balance for this process based on 100kg of gas burned.

## 1.7 BRIEF OUTLINE ABOUT THE WORK

☞Distinguish between classical and industrial chemistry

☞Classify the chemical industry in terms of scale, raw materials, end use and value addition

☞Distinguish between unit operations and unit processes

☞Describe chemical processes by means of flow diagrams

☞Carry out material balances for a simple process

## 1.8 SUMMARY

This learning activity introduces you to industrial chemistry and the chemical indus-try and enables you to study subsequent units more easily. It includes the following topics: Introduction to industrial chemistry, classification of the chemical industry, raw materials for the chemical industry, unit operations and unit processes, flow diagrams, material and energy balances. The various readings given supplement the material presented in this module. At the end of the unit, there are exercises you are required to do to test your understanding of the unit.

## 1.9 REFERENCES

1. Chang R. (1991). Chemistry, 4$^{th}$ Edition, McGraw-Hill Inc. New York.

2. Chang R. and Tikkanen W. (1988). The Top Fifty Industrial Chemicals.

3. Price R.F. and Regester M.M. (2000), WEFA Industrial Monitor, 2000-2001, John Wiley & Sons Inc., New York.

# CHAPTER – 2

## Unit Operations and Unit Processes

☞CONTENTS

**2.1 Unit Operations**

In Chapter 1, we defined unit operations as physical transformations. They are very many and include size reduction, size enlargement and separation of mixtures. In this unit, we shall look at operation principles of equipment in these unit operations and their application in the chemical industry.

## 2.1.1 Size Redtiucon

Size reduction refers to all the ways in which particles are cut or broken into smaller pieces. The objective is to produce small particles from big ones for any of the following reasons:

1. To reduce chunks of raw materials to workable sizes e.g. crushing of mineral ore.

2. To increase the reactivity of materials by increasing the surface area.

3. To release valuable substances so that they can be separated from unwanted material.

4. To reduce the bulk of fibrous materials for easier handling.

5. To meet standard specifications on size and shape.

6. To increase particles in number for the purpose of selling.

7. To improve blending efficiency of formulations, composites e.g. insecticides, dyes, paints

### 2.1.1.1. Principles of size reduction

Most size reduction machines are based on mechanical compression or impact.

When a solid is held between two planes and pressure is applied on one plane, the solid is fractured and breaks into fragments when pressure is removed. The fragments formed are of different sizes. An example of an industrial equipment that is based on compression is a jaw crusher. Impact is the breaking up of material when it is hit by an object moving at high speed. The product contain coarse and fine particles. A ball mill is based on impact.

## 2.1.1.2. Jaw Crusher

Fig.2.1 is a schematic diagram of a jaw crusher.

**Fig 2.1** Jaw crusher

A jaw crusher consists of a vertical fixed jaw and another swinging jaw that moves in the horizontal plane. In the diagram above, the jaws are coloured red. The two jaws make 20-30° angle between them. The swinging jaw closes about 250 to 400 times/min. Feed is admitted between the jaws. It is crushed several times between the jaws before it is discharged at the bottom opening.

A jaw crusher produces a coarse product.

## 2.1.1.3. Ball Mill

A ball mill is a tumbling mill generally used for previously crushed materials. It is generally used to grind material 6mm and finer, down to a particle size of 20 to 75 microns.

**Fig. 2.2** Ball mill

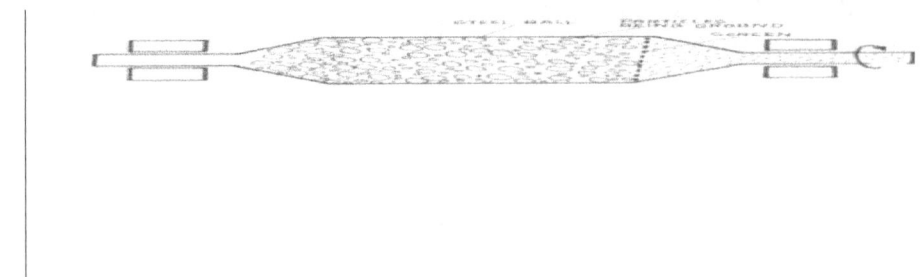

The operation of a ball mill is illustrated in Fig 2.2. The mill consists of a cylinder containing a mixture of large and small steel grinding balls and the feed. When the cylinder is rotated, the rotation causes the balls to fall back into the cylinder and onto the material to be ground. The rotation is usually between 4 to 20 revolutions per minute, depending on the diameter of the mill. The larger the diameter, the slower the rotation. If the speed of the mill is too great, it begins to act like a centrifuge and the balls do not fall back, but stay on the perimeter of the mill. The point where the mill becomes a centrifuge is called the critical speed. Ball mills usually operate at 65% to 75% of the critical speed. A ball mill is suitable for dry- or wet- milling of various material in cement, fertilizer, metallurgical industries and other industries. Fig 2.3 is a ball mill installed in acement factory.

## 2.1.2 Size Enlargement (Agglomeration)

Size enlargement, also referred to as agglomeration, is carried out when particles are too small for use in a later stage of the process. For example in metal extraction, some particles may be too fine to be fed into a blast furnace.

### 2.1.2.1. Purposes of size enlargement

The following are some of the purposes of size enlargement in various industries:

1. Reduce dusting losses
2. Reduce handling hazards particularly with respect to irritating and obnoxious powders.
3. Render particles free flowing.
4. Densify materials.
5. Prevent caking and lump formation
6. Provide definite quantity of units suitable for metering, dispensing and administering

7.  Produce useful structural forms

8.  Create uniform blends of solids which do not segregate

9.  Improve appearance of products

10. Permit control over properties of finely divided solids e.g. solubility, porosity, surface volume ratio, heat transfer

11. Separate multicomponent particle size mixtures by selective wetting and agglomeration

12. Remove particles from liquids

In size enlargement, small particles are gathered into larger, relatively permanent masses in which the original particles can still be identified. The products of size enlargement are either regular shapes e.g. bricks, tiles, tablets, pellets or irregular shapes such as sintered ore.

Agglomerators are used to increase the particle size of powders. There are two basic types of agglomerators; compaction and non-compaction agglomerators. The compaction type uses mechanical pressure (and often very high pressures) to "press" the powders together. For these, binders are sometimes not needed to make the particle. Pellet mills are compaction agglomerators.

### 2.1.2.2. Pellet mills

Moist feed in plastic state is passed through a die containing holes. The die is supplied with power to rotate around a freely rotating roller. The friction of material in the die holes supplies resistance necessary for compaction. A knife cuts the exudates into pellets. This is shown in Fig. 2.4. Bonding agents such as glue or starch may be mixed with the feed.

Pellet quality and capacity depends on:

- Feed properties e.g. moisture

- Lubricating characteristics

- Particle size

- Abrasiveness

Die characteristics and speed

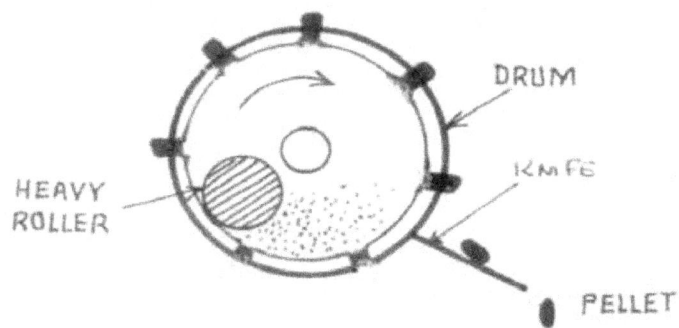

**Fig 2.4**     The pellet mill

A picture of a pellet mill converting wood planings and sawdust into fuel pellets is shown in Fig 2.5. These raw materials are compressed under high pressure into small, cylindrical rolls. Pellets gain their firmness solely from the pressing process without addition of any chemical or synthetic adhesive agent.

**Fig. 2.5** Photograph of a pelletizer in operation that converts planings into fuel pellets.

### 2.1.2.3. Tumbling agglomerators

The common action of most non-compaction agglomerators is to keep the powders in motion by tumbling, vibrating or shaking, while spraying a correct amount of liquid binder. The binder is specially selected to hold the smaller particles together, creating a larger particle. After the particles stick together to form a nucleus or germ, then follows the layering or deposition of layers of the raw materials into previously formed nucleus. This requires high recycle ratio whose increase leads to larger and denser agglomerates of high wet strength. It also requires low moisture content in spite of the fact that increase in liquid content leads to increase in agglomerate size. The layering process is shown in Fig. 2.6

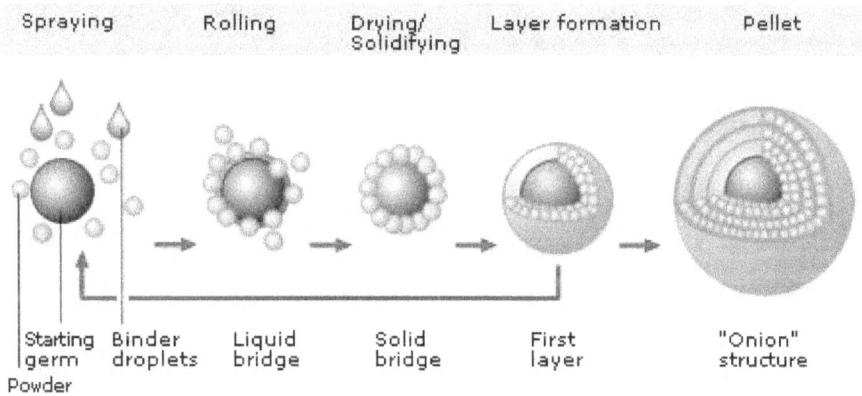

**Fig. 2.6** Illustration of the powder layering process.

The formed agglomerates are subject to the following forces:

a)    Destructive forces within the feed as particles impact on each other during the rolling action

b)    Cohesive forces holding pellets together

Optimum agglomeration is obtained when correct tumbling and cascading motion occurs in the charge. Motion is caused by centrifugal forces. The devices may be operated at an angle.

Two types of tumbling agglomerators are used: inclined pan agglomerator and a drum agglomerator.

### 2.1.2.3.1   Inclined pan agglomerator

This is shown in Fig 2.7. It consists of pan rotating at an incline. It is fed with the powdery raw material. Material layers over a nucleus particle to form balls. Enlarged balls roll off the pan. Fine materials silts down through the large balls and remain in the pan.

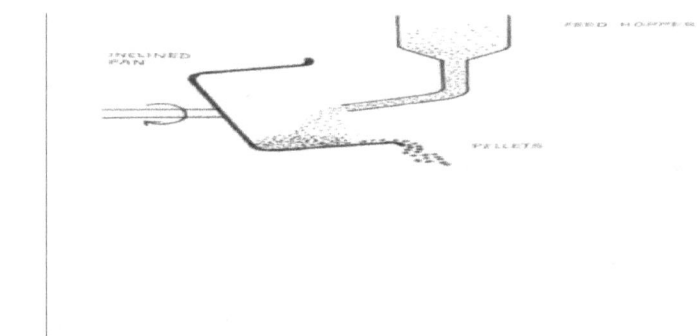

**Fig 2.7** Inclined pan agglomerator

The following are the advantages of an inclined pan:

1) Uniform product without need for a screen

2) Low equipment cost which is simple to control

3) Easy observation of the balling action

However, an inclined pan has one disadvantage: it produces dust.

## 2.1.2.3.2. Drum agglomerator

Fig 2.8 is an illustration of the operation of a drum agglomerator.

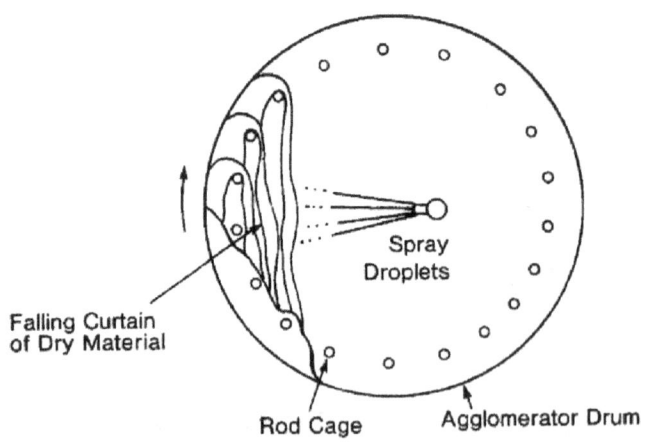

**Fig 2.8** Cross sectional view of an agglomerator

As the drum rotates clockwise, the bars (labeled "Rod Cage") lift the finer powders and create a falling-curtain of the smaller and dry particles. The liquid binder is sprayed ("Spray Droplets") onto this curtain, which preferentially agglomerate only the small particles lifted by the Rod Cage.

A drum agglomerator has the following advantages over a pan agglomerator:

1)    Large capacity

2)    Large retention time if required

3)    Less sensitivity to upsets in the system due to the dumping effect of large recirculating load

The drum agglomerator has one disadvantage. Because particles of various sizes are produced, a screen is required to separate enlarged particles from the smaller particles.

## 2.1.3.    Separation Of Materials

In this section, we will learn how differences in the physical properties of materials are used to separate mixtures in the chemical industries.

### 2.1.3.1. Magnetic Separation

If a mixture containing magnetic materials and non-magnetic materials is subjected to a magnetic field, there is competition for the particles between several forces namely, magnetic, inertia, gravitational and interparticle forces.

Three products can be obtained during magnetic separation. These are:

•    A strongly magnetic product

•    A weakly magnetic (middlings) product

- A non-magnetic (tailings) product

Separation is carried out either dry using belt lifting magnets or wet using drum magnetic separators. This technology is applied in mineral ore processing, as we shall see in Unit 4.

The method used for dry particles is illustrated schematically in Fig 2.9.

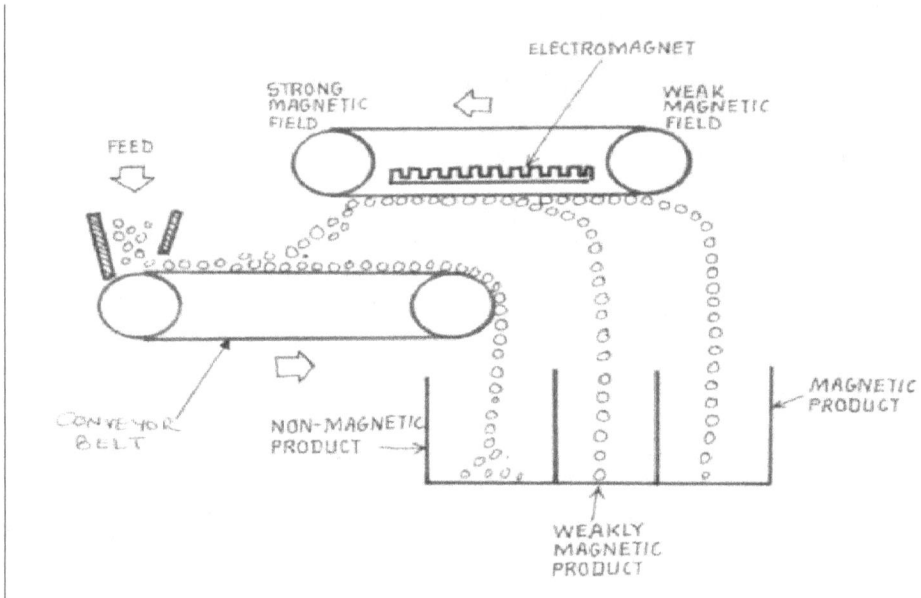

**Fig 2.9**    Illustration of the principle of dry magnetic separation

Material to be separated is fed into the first conveyor. Above this conveyor is another conveyor with an electromagnet inside. The electromagnetic field decreases towards the right. Strongly and weakly magnetic materials are attracted and picked by the magnet. The non-magnetic materials continue to be conveyed by the bottom conveyor and drop in the first bin. As the strength of the electromagnet weakens towards the right, the middlings i.e. the weakly magnetic materials lose attachment and drop in the middle bin. The strongly magnetic materials drop off at the end of the electromagnet into the third bin.

## 2.1.3.2. Froth Flotation

This is a process in solids-liquids separation technology that uses differences in wettability of various materials such as mineral ores. Although these materials are generally hydrophilic, the surface properties of components they contain may vary within a very narrow range. These small differences can be amplified by selective adsorption that makes some of the particles hydrophobic. Such hydrophobic particles in a water suspension are floated by attaching them to air bubbles.

**Making particles hydrophobic and floatable**

A special surface-active agent (surfactant) called **collector** or **promoter** is added to the suspension. Collectors are usually $C_2$ to $C_6$ compounds containing polar groups. They include fatty acids, fatty acid amines and sulphonates among others. Collector selection depends on the material being separated. The collector molecule adsorbs on to the solid surface via the polar (charged) group. This reaction is known as chemisorption. The hydrocarbon chain is facing the aqueous phase. This is shown in Fig.2.10.

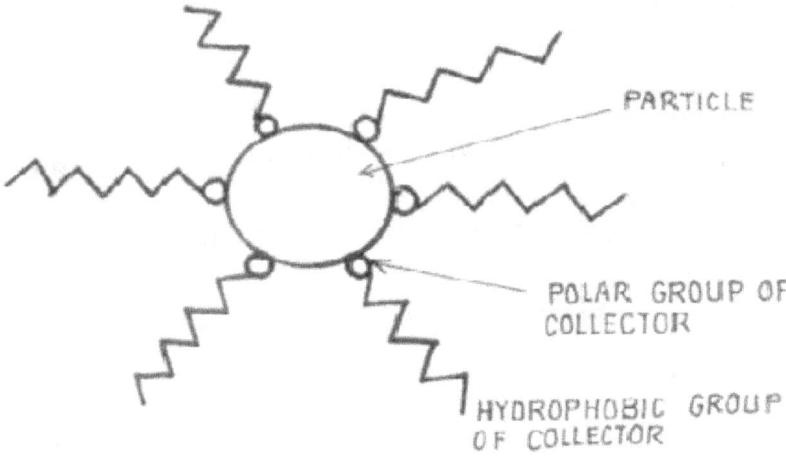

PARTICLE

POLAR GROUP OF COLLECTOR

HYDROPHOBIC GROUP OF COLLECTOR

**Fig 2.10** how a collector renders a particle hydrophobic

A layer probably, a monolayer of the collector molecules become attached to the surface of the particle. Because the hydrocarbon chain and the water do not mix, the coated particle surface becomes hydrophobic. By being hydrophobic, a particle repels water. This results in the weakening of the forces acting between the particle surface and water and hence the diminishing of surface-water interactions at solid-surface interface. This causes the displacement of water film from the wetted solid surface by air. In addition to the use of collectors to change the surface property of the particles, other chemicals may be added to further modify either the particles to be floated, or the particles that are to remain in the suspension. Such chemical subs-tances are called modifiers.

### 2.1.3.2.2 Flotation cell

The flotation cell is shown in Fig. 2.11. The material is ground in water to a maximum 250μm. It is introduced into the flotation cell. A frothing agent is added to create a generous supply of fine bubbles when air is sparged. Examples of frothers include pine oil and methyl amyl alcohol. The collector and other additives are added. Hydrophobic particles are collected at the air-bubble interface. The bubbles with attached mineral particles rise to the surface where the material is removed. Particles that are readily wetted by water (hydrophilic) tend to remain in the water suspension.

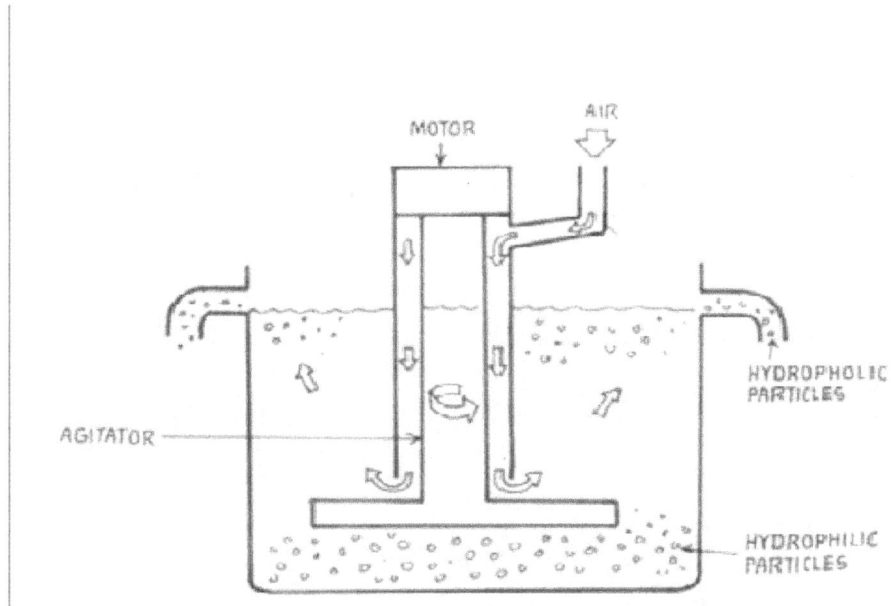

**Fig 2.11**     a flotation cell

## 2.1.3.3.    Fractional Distillation

Distillation is used to separate a mixture of miscible liquids which have different volatilities. Suppose a mixture with low concentration of the more volatile component is distilled and the vapour condensed. The condensate which we refer to as distillate will be more concentrated with this component than the feed. If we return the distillate to the distillation apparatus and distill it to a second distillate, this distillate will be more richer in the more volatile component than the first distillate. If we continue this process, we will approach a pure distillate of the more volatile component. The greater the relative volatility between the two components, the fewer the needed distillation stages. This is the concept of fractional distillation. It is used when:

- Boiling points of mixture components are close
- Volatilities of the components are close

This is the case in petroleum refining.

Industrially, fractional distillation is carried out in distillation columns also known as distillation towers. They are like many distillation stills stack together vertically.

The fractionation columns can be batch or continuous and they can be many in series. Fig. 2.13 is a picture of an industrial distlillation plant.

**Fig 2.13**    An industrial fractionating plant

For distillation to take place in a distillation column, both vapour and liquid flowing up and down respectively must be brought into intimate contact. This is done by packing the column with inert solids, or installing plates at regular intervals throu-ghout the column height. Small distillation columns are normally packed while large distillation columns are plated.

In plated columns, we need to provide for both vapour path and liquid paths at each plate. The plates are perforated and the vapour passes through the perforations. The liquid flows through pipes known as downcomers next to the colum wall.

In Fig 2.13, the components of a continuous distillation column are shown.

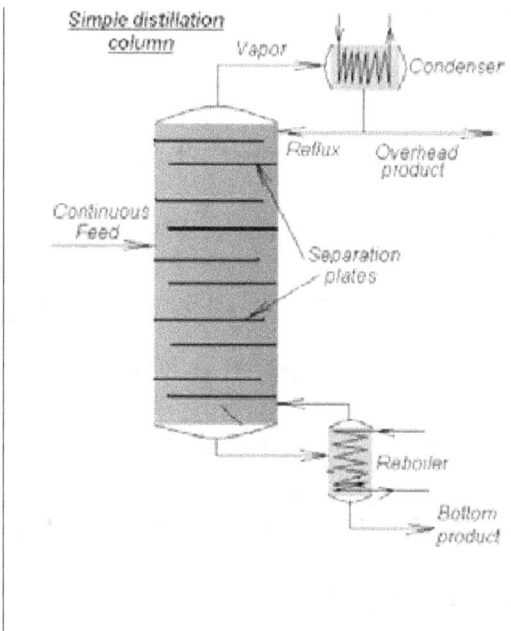

**Fig 2.13**    A continuous distillation column

**Rectification and stripping in a distillation column**

Let us look at what happens inside a distillation column when it is running. Suppose there are two components A and B being separated with A being more volatile than B. Component A leaves at the top. At each distillation plate, the liquid mixture is at boiling point. The boiling point of A, $T_A$ is low compared to $T_B$, the boiling point of B coming out at the bottom. We can therefore conclude that the inside of the column becomes colder upwards.

Vapour is generated at the reboiler and it rises up from the bottom of the column. At each distillation stage or plate, a relatively hotter vapour contacts a cooler liquid coming down the column. Some of the vapour condenses and the resulting conden-sate has more of the less volatile component B, thus resulting in a vapour rcher in A. Simultaneously, some of the liquid picks the latent heat generated by the condensing vapour and vapourizes. The vapourizedportion contain more of the more volatile component A and therefore, the liquid leaving the plate is depleted of A and enriched with B. This process is repeated up the column. After the topmost vapour is condensed at the condenser, some of the distillate is returned to the column at the top plate. The returned liquid is called reflux. Enriching the vapour with the more volatile component above the feed location is known as **rectification.** The removal or depletion of this component from the liquid below the feed location is known as **stripping**.

### 2.1.3. Other Unit Operations

There are many unit operations that are employed in the chemical industry. It is impossible to cover all of them in this unit. In Table 2.1 a summary of some other unit operations is given.

**Table 2.1 Other unit operations**

| Unit Operation | Purpose | Application |
|---|---|---|
| Electrostatic separation | Separation of solids on the basis of the difference in electrical conductivity of components | Mineral ore dressing |
| Sedimentation | Separation of solids from | Water treatment plants |

| | liquids mostly by gravitational forces | |
|---|---|---|
| Crystallization | Separation of solid particles from their saturated solutions | Sugar manufacture |
| Solid–liquid extraction or leaching | Extraction of a soluble solid from its mixture with an inert solid, by use of liquid solvent in which it is soluble. | Mineral ore processing |
| Spray drying | A liquid containing a dissolved solid is sprayed and contacted with hot air which evaporates the solvent yielding a powdered product | Production of pigments, detergent powder, powdered milk, synthetic resins and inorganic salts |
| Liquid – liquid or solvent extraction | Separation of a liquid solute from its mixture with another ''inert' liquid by means of another liquid in which it is soluble | Solvent recovery, removal of naphthenic and aromatic compounds from lubricating oil |
| Absorption | Removal of a component from a gas mixture by dissolving it in a liquid | removal of hydrogen sulphide ($H_2S$) from hydrocarbon gases using alkaline solutions. |

## 2.2. Unit Processes

In Unit 1, we defined unit processes as chemical transformations or conversions. Unit processes are the core of industrial synthetic chemistry and are dominant in organic processes. We will look at some of the unit processes that we are likely to encounter in subsequent learning activities.

### 2.2.1. Polymerisation

The term polymer comes from two Greek words: "polys" which

means "many" and "meros" which means "parts." A polymer is therefore a substance having hundreds or thousands of many smal identical parts known as monomers, which are bonded together covalently in a chemical process known as polymerization.

## 2.2.1.1. Polymerization reactions

For polymerisation to yield polymers with long chain (high polymers), the monomers must:

- Be polyfunctional i.e. contain at least two reactive groups

- Not give cyclic products by intramolecular ring closure because this will terminate polymerisation.

Polymerization reactions fall into two general classes:

1. Addition or chain polymerization involving successive stages of reaction **initia-tion, propagation** and **termination.** Examples of addition polymers includepolyethylene, polypropylene, polyvinyl chloride and polystyrene.

2. Condensation or step-reaction polymerization. This involves condensation reaction between two polyfunctional molecules, sometimes with the elimina-tion of a small molecule such as water. Nylon is a condensation polymer of hexamethylenediamine and adipic acid as shown in the equation below:

$H_2N-(CH_2)_6-NH_2 + HOOC-(CH_2)_4-COOH \longrightarrow H_2O + -(NH-(CH_2)_6-NH-CO-(CH_2)_4)-$

If one of the reactants in a step-reaction polymerization contain more than two func-tional groups, a cross-linked polymer is obtained.

## 2.2.1.2 Free radical polymerization

One of the most common and useful reactions for making polymers is free radical polymerization. It is used to make polymers from vinyl monomers, that is, from small molecules containing carbon-carbon double bonds. Polymers made by free radical polymerization include polystyrene, polymethylmethacrylate, polyvinyl acetate and branched polyethylene

**Initiation:** The whole process starts off with a molecule called an initiator. This isa molecule like benzoyl peroxide or 2,2'-azo-*bis*-isobutyrylnitrile (AIBN). The special characteristic of these molecules is that they have an ability to split in unusual way. When they split, the pair of electrons in the bond, which is broken, separates to produce two initiator fragments, each of which has one unpaired electron. Molecules like this, with unpaired electrons are called *free radicals*

The carbon-carbon double bond in a vinyl monomer, like ethylene, has a pair of electrons, which is very easily attacked by the free radical. When this happens, a new chemical bond is formed between the initiator fragment and one of the double bond carbons of the monomer molecule. This electron, having nowhere else to go, associates itself with the carbon atom, which is not bonded to the initiator fragment. This whole process, the breakdown of the initiator molecule to form radicals, followed by the radical's reaction with a monomer molecule is called the *initiation* step of the polymerization.

**Propagation:** This new radical reacts with another ethylene molecule in the exactsame way as the initiator fragment did. A free radical is formed when this reaction takes place over and over again and the chain grows. The adding of more and more monomer molecules to the growing chain is called *propagation*.

Because we keep remaking the radical, we can keep adding more and more ethylene molecules, and build a long chain. Self-perpetuating reactions like this are called *chain reactions*.

**Termination:** Termination is the third and final step of a chain-growth polymerization. Free radicals are unstable, and eventually they find a way to become paired without generating a new radical. Then the chain reaction will grind to a halt. This happens in several ways. The simplest way is for two growing chain ends to find each other. The two unpaired electrons then join to form a pair, and a new chemical bond joining their respective chains. This is called *coupling*. Coupling is one of two main types of *termination reaction*. Another way in which the unpaired electrons can shut down polymerization is called *disproportionation*. In disproportionation, the unpaired electron of one chain finds an electron in the carbon-hydrogen bond of the next carbon atom forming a double bond at the end of the polymer chain.

Sometimes, the unpaired electron at the end of a growing chain pairs with an electron from a carbon-hydrogen bond along the backbone of another polymer chain. This starts a new chain growing out of the middle of the main chain. This is called *chaintransfer to polymer*, and the result is a *branched polymer*. It is especially a problemwith polyethylene, so much that linear non-branched polyethylene cannot be made by free radical polymerization.

Polymerisation products are numerous with many uses and include phenolic resins, alkyl resins, polyamides, polyesters, elastic foams, silicon polymers, isocyanate polymers, epoxy resins, adhesives, coatings, polyethylene, vinyl polymers and acrylic polymers (for paint industry) to mention but a few.

### 2.2.1.3. Emulsion polymerization

Emulsion polymerization is a type of free radical polymerization that usually starts with an emulsion consisting of water, monomer and surfactant. The most common type of emulsion polymerization is an oil-in-water emulsion, in which droplets of monomer (the oil) are emulsified (with surfactants) in a continuous water phase. Water-soluble polymers, such as certain polyvinyl alcohols or hydroxyethyl celluloses, can also be used to act as emulsifiers/stabilizers. Emulsion polymerization is used to manufacture several commercially important polymers. Many of these polymers are used as solid materials and must be isolated from the aqueous dispersion after polymerization. In other cases, the dispersion itself is the end product. A dispersion resulting from emulsion polymerization is often called a latex (especially if derived from a synthetic rubber) or an emulsion (even though "emulsion" strictly speaking refers to a dispersion of a liquid in water). These emulsions find applications in ad-hesives, paints, paper and textile coatings. Because they are not solvent-based, they are eco-friendly.

Advantages of emulsion polymerization include:

•    High molecular weight polymers can be made at fast

polymerization rates

- The continuous water phase is an excellent conductor of heat and allows the heat to be removed from the system, allowing many reaction methods to increase their rate.

- Since polymer molecules are contained within the particles, viscosity remains close to that of water and is not dependent on molecular weight

- The final product can be used as is and does not generally need to be altered or processed.

Disadvantages of emulsion polymerization include:

- Surfactants and other polymerization adjuvants remain in the polymer or are difficult to remove

- For dry (isolated) polymers, water removal is an energy-intensive process

Emulsion polymerizations are usually designed to operate at high conversion of monomer to polymer. This can result in significant chain transfer to polymer

The Smith-Ewart-Harkins theory for the mechanism of free-radical emulsion polymerization is summarized by the following steps:

- A monomer is dispersed or emulsified in a solution of surfactant and water forming relatively large droplets of monomer in water.

- Excess surfactant creates micelles in the water.

- Small amounts of monomer diffuse through the water to the micelle.

- A water-soluble initiator is introduced into the water phase where it reacts with monomer in the micelles. This is considered Smith-Ewart Interval 1.

- The total surface area of the micelles is much greater than the total surface area of the fewer, larger monomer droplets; therefore the initiator typically reacts in the micelle and not the monomer droplet.

- Monomer in the micelle quickly polymerizes and the growing chain terminates. At this point the monomer-swollen micelle has turned into a polymer particle. When both monomer droplets and polymer particles are present in the system, this is considered Smith-Ewart Interval 2.

- More monomer from the droplets diffuses to the growing particle, where more initiators will eventually react.

- Eventually the free monomer droplets disappear and all remaining monomer is located in the particles. This is considered Smith-Ewart Interval 3.

- Depending on the particular product and monomer, additional monomer and initiator may be continuously and slowly added to maintain their levels in the system as the particles grow.

- The final product is a dispersion of polymer particles in water. It can also be known as a polymer colloid, a latex, or commonly and inaccurately as an 'emulsion'.

Emulsion polymerizations have been used in batch, semi-batch, and continuous pro-cesses. The choice depends on the properties desired in the final polymer or dispersion and on the economics of the product.

## 2.2.2.    Alkylation

Alkylation is the introduction of an alkyl radical by substitution or addition into an organic compound. For example, the combining of an olefin to a hydrocarbon is an alkylation reaction. In the presence of an acid catalyst such as hydrogen fluoride or sulphuric acid, this reaction is used for the conversion of gaseous hydrocarbons to gasoline. The processes are usually exothermic and similar to polymerisation. Another example is the formation of 2,2-Dimethylbutane from ethylene and isobutane:

Alkylation reactions include the binding of an alkyl group to:

1)    Carbon (to make products such as gasoline alkylates, ethylbenzeneetc)

2)    Oxygen of a hydroxyl group of an alcohol or phenol (ethers, alkaloids)

3)    Trivalent nitrogen (amines)

4)    A tertiary nitrogen compound (quaternary ammonium compounds)

5)    Metals

6)    Miscellaneous elements such as sulphur or silicon.

In 6), the alkyl is in the form of alkyl halide or ester.

Apart from gasoline, other classes of products from alkylation reactions include pharmaceuticals, detergents, disinfectants, dyes and plastics.

Alkylates of active methylenes are easily prepared using a base such as ethoxide, $EtO^-$

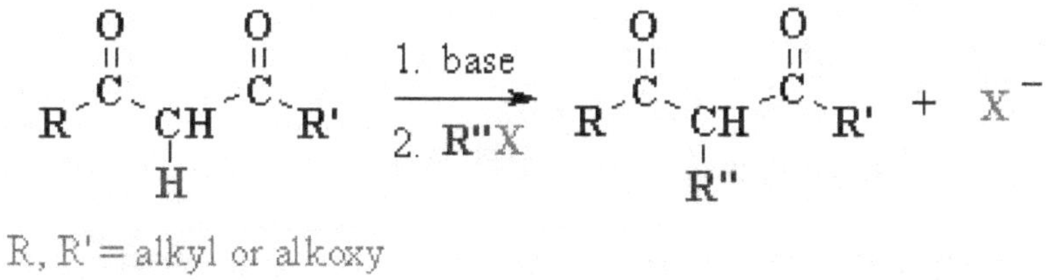

R, R' = alkyl or alkoxy

Methyl and primary halides are most suitable for alkylation reactions. In principle both of the hydrogens can be replaced with alkyl groups:

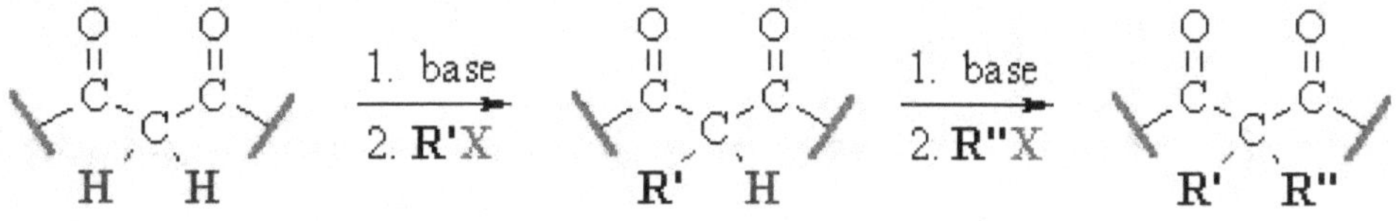

This can be utilized to form a cyclic system by using a dihalide as shown below:

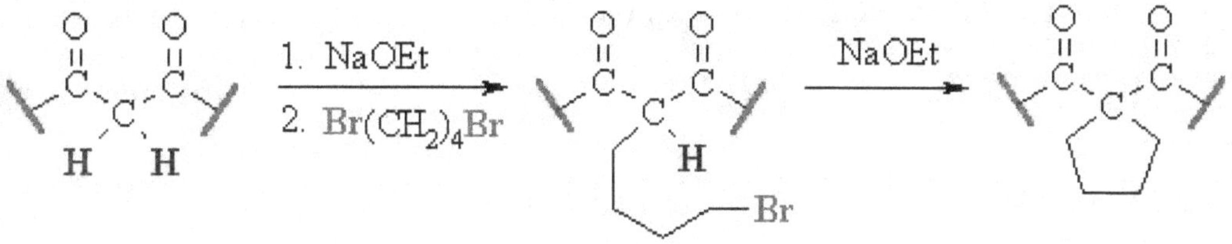

## 2.2.3.    Hydrolysis

In the hydrolysis of either organic or inorganic compounds, water and another compound undergo double decomposition to form two products. The hydrogen from the water goes to one product while the hydroxyl goes to the other product as illustrated in the following equation:

$$XY + H_2O \longrightarrow HY + XOH$$

If XY were an inorganic compound, this would be the reverse of neutralization. But in organic chemistry, hydrolysis has a wider scope, which includes:

- Inversion of sugar

- Breaking down of proteins

- Saponification of fats and oils

These reactions can be carried out with water alone. However, there are agents that accelerate or catalyse the hydrolysis. These include:

- Alkalis

- Acids

- Enzymes

## 2.2.3.1.  Hydrolysis of esters

This hydrolysis is referred to as saponification. A good example is the saponification of fats and oils to glycerol and either soap or fatty acids. Ester hydrolysis is reversible and is catalysed by both the oxonium ion ($H_3O^+$), and hydroxyl ion ($OH^-$). That is, it can be either acid or alkali catalysed. Addition of acid speeds up reaction without shifting the equilibrium significantly. On the other hand, alkali addition accelerates reaction and shifts the reaction to the right so that it goes to completion.

## 2.2.4. Other Unit Processes

In table 2.2, you will find other unit processes with their definition and industrial application.

**Table 2.2. Other unit processes with their industrial applications.**

| Unit process | Brief description | Industrial applications |
|---|---|---|
| Sulphonation | A chemical process that involves introduction of sulphonic acid group (SO$_2$OH) or its corresponding salt or sulphonyl halide (=SO$_2$Cl) into an organic molecule. The sulphonating agents include sulphuric acid (98%), sulphur trioxide in water (oleum) and fuming sulphuric acid. | Intermediates in manufacture of phenol, xylene, dodecyl benzene sulphonic acid detergent, polystyrene, naphthalene derivatives and aliphatic sulphonated compounds. |
| Esterification | A chemical process in which an ester and water are formed when an organic radical is substituted for in a molecule by an ionisable hydrogen of an acid. | Production of synthetic fibres like polyethylene terephthalate, manufacture of alkyl resins and polyvinyl acetate, preparation of terpene and cellulose esters |
| Hydrogenation | A chemical reaction of molecular hydrogen with another substance in the presence of a catalyst. | Manufacture ammonia (see details in learning activity two), manufacture of liquid fuels, hydrogenated vegetable fats, hydrogenation of carbohydrates to propylene, glycol and sorbitol and many others. |
| Halogenation (chlorination, bromination and iodination) | Involves addition of one or more halogen atoms to an organic compound | Chlorinated compounds are used in the chlorohydrocarbons such as chloroform, ethylene chlorohydrin (freon), DDT, carbon tetrachloride, olefinic acids, acid chlorides, etc. |
| Nitration | This is the introduction of one or more nitro groups (-NO$_2$) into an organic compound. Monovalent atoms or groups of atoms are replaced by the nitro group | Industrial solvents, dyestuffs, explosives, pharmaceuticals and as intermediates in the production of amines |

# CHAPTER – 3

## Inorganic Extractive Metallurgy

### ☞CONTENTS

3.4.5 Manufacture of Pig Iron

3.4.6 Reactions of the blast furnace

## 3.5   Extractive Metallurgy Of Aluminium

3.5.1 Chemical treatment of bauxite

3.5.2 Reduction of aluminum from aluminium oxide

## 3.6 Extractive Metallurgy Of Copper

3.6.1 Concentrating

3.6.2 Roasting

3.6.3 Matte smelting

3.6.4 Fire Refining

3.6.5 Electrolytic refining

## 3.1. Mineral ores

An ore is a mineral deposit which can be profitably exploited. It may contain three groups of minerals namely:

a)   valuable minerals of the metal which is being sought

b)   compounds of associated metals which may be of secondary value

c)   gangue minerals of minimum value.

Almost all metals are derived from mineral ores. There are also ores that contain non-metals such as sulphur. Generally, the valuable mineral in an ore may be found in the form of native metal, oxides, oxy-salts, sulphides or arsenides.

During mining, large open pits are excavated by breaking the ore using explosives. Ores as mined may be in large lumps and therefore, some size reduction is done at the mine. The ore is shoveled into trucks and transported to the factory. If the mineral ore is found in waterbeds, mining is carried out by dredging. For example, sand is dredged from river beds.

## 3.2. Ore dressing

Before the ores are subjected to the main chemical treatment steps, they are pre-trea-ted by a series of relatively cheap processes, mainly physical rather than chemical in nature. These processes constitute what is known as **ore dressing.** They are meant to effect the concentration of the valuable minerals and to render the enriched material into the most suitable physical condition for subsequent operations. Ore dressing may include:

*   **Size Reduction** to such a size as will release or expose all valuable mine-rals
*   **Sorting** to separate particles of ore minerals from gangue (non-valuable)minerals or different ores from one another
*   **Agglomeration** may be carried out sometimes before a roasting operation

If the ores are rich in the valuable mineral, above processes may not add value. Such ores can be ground, sized and blended with other ores in order to provide a homoge-neous feed to say, a blast furnace or reaction bed.

## 3.2.1.    Size Reduction

Size reduction may be carried out by first crushing the ore down to 7mm maximum followed by grinding to smaller sizes. Jaw crushers can be used deep in the mine to prepare the ore for transportation to the surface e.g. using bucket elevators.

## 3.2.2. Sizing

Screens are used to separate particles according to size and may not affect the concentrations of the ore minerals. Particles are separated into oversize and undersize.

## 3.2.3. Sorting

The particles may be sorted by classification, flotation or magnetic methods. Classifiers

These are devices that separate particles according to their different rates of travel under gravity through a fluid medium such as water. Particles of different densities, sizes and shapes have different falling velocities. Classifiers include rake classifiers and jigs.

## 3.2.4. Flotation

Flotation uses difference in surface properties of the individual minerals. It is readily applied to very fine concentrates and can distinguish ore mineral from gangue, and also, one ore mineral from another.

## 3.2.5. Magnetic Separation

Ferromagnetic magnetite or iron minerals which can be chemically altered to produce magnetite may be sorted out using a magnetic separator as described in Unit 2.

## 3.2.6. Electrostatic Separation

Minerals have a wide range of electrical conductivity and can be distinguished by this property. If several kinds of particles are given an electrostatic charge and are then brought into contact with an electrical conductor at earth potential, the charge will leak away from good conductors much more rapidly than from poor conductors. While the charge remains, the particle will cling to the conductor by electrostatic attraction. The weakly conducting minerals will therefore remain attached to the conductor longer than the good conductors, so affording a means of separating minerals whose conductivities differ appreciably. Electrostatic separators operate on thin layers of material. The principle is illustrated in Fig. 3.1.

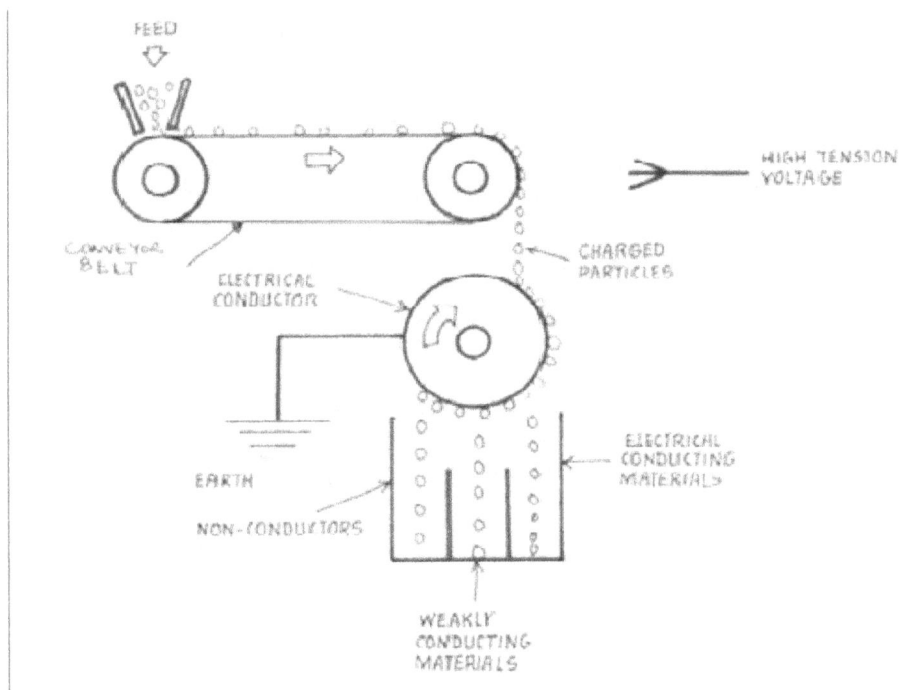

**Fig 3.1**    Electrostatic separation

## 3.2.7.    Dewatering and filtration

After sorting and leaching, it is necessary to separate the solid and liquid phases.

Coarse solids may be freed from most of their moisture by draining. Slurries with particles which can settle may be separated from the bulk of the liquid by settling and subsequent decantation. These dewatering methods may reduce moisture content to

50%. The moisture content may be reduced further by filtration and drying. If the va-luable ore is in the filtrate, it can be recovered by evaporation followed by drying.

### 3.2.8. Agglomeration

When a particle size of an ore or concentrate is too small for use in a later stage of treatment e.g. in a blast furnace, it must be reformed into lumps of appropriate size and strength. This is done by any of the following methods:

- pelletizing
- briqueting
- sintering

### *Briqueting*

This is a mechanical process of agglomeration in which the materials, after mixing with water and necessary bonding agents are pressed or extruded into brick or block form. These blocks are then dried and hardened by heating. Use of hydraulic cement allows hardening to be carried out cold. Briqueting is not popular in mineral ore agglomeration.

### *Sintering*

Sintering involves diffusion of material between particles. It is applied to the consolidation of metallic and ceramic powder compacts which are heated to temperatures approaching their melting points to allow diffusion to take place at the points of contact of particles so that they grow together to form a rigid entity. The process can be envisaged as a net migration of vacancies into the solid at the highly curved energy surfaces near points of contact and again at low energy areas away from contact points

Sintering may be accompanied by a chemical reaction.

## 3.3.    Extraction Processes

So far we have been dealing with unit operations that prepare the ore for chemical reactions used to extract the valuable metal from the ore. Now we want to look at extraction and refining of the metal.

## 3.3.1.    Calcination

This is the thermal treatment of an ore to effect its decomposition and the elimination of a volatile product, usually carbon dioxide or water. The following are calcinations reactions

$$CaCO_3 \rightarrow CaO + CO_2 \qquad T = 1000^0C$$

$$MgCO_3 \rightarrow MgO + CO_2 \qquad T = 417^0C$$

$$MnCO_3 \rightarrow MnO + CO_2 \qquad T = 377^0C$$

$$FeCO_3 \rightarrow FeO + CO_2 \qquad T = 400^0C$$

Calcination may be carried out in rotating kilns using countercurrent flow for efficient heat transfer.

## 3.3.2. Roasting

Roasting involves chemical changes other than decomposition, usually with furnace atmosphere. A roast may effect calcinations and drying as shown below.

$$2CuS + O_2 \rightarrow Cu_2S + SO_2 \text{ (calcination)}$$

$$Cu_2S + O_2 \rightarrow 2Cu + SO_2 \text{ (roasting)}$$

### 3.3.3.Smelting

This is essentially a smelting process in which the components of the charge in the molten state separate into two or more layers which may be slag, matte, speiss or metal

- matte: heavy sulphide material
- slag: light oxide material
- speiss: iron oxide, insoluble in matte, slag or metal; it may contain elements

Smelting of metal involves reduction, usually by carbon or coal or coke and may be performed in a blast furnace or an electric furnace.

In the blast furnace, coke is burned into $CO_2$ which reacts further with the carbon to form CO. The ascending gases pre-heat the solid charge descending the stack and reduce metal oxides to metal. This then is a process of drying followed by calcination and roasting. The metal melts and the slag forms gangue and flux.

Where fusion or reduction temperature is above $1500^0C$, electric melting is most appropriately applied.

### 3.3.4. Refining

Electrolysis may be used for metal extraction and metal refining. In fire refining, extracted metals are brought into liquid state and their composition finally adjusted.

In some case, this may be simple smelting to allow cathodically entrained hydrogen to escape by diffusion. In other cases, impurities may react to form compounds which are insoluble in the molten state. Converters are used for oxidizing impurities out of blast furnace iron in steel-making and for oxidation of sulphur from copper and nickel matte. Distillation may also be applied in metal purification.

## 3.4. Extractive Metallurgy Of Iron

### 3.4.1.   Uses of iron

Iron is used in the forms shown below as material of construction for machines, plants, buildings, locomotives, ships, automobiles, railway lines and for many other things. All these forms are obtained from pig iron which is first obtained from the iron ore.

a.   White cast iron obtained when molten low silicon, high manganese pig iron is rapidly cooled.

b.   Grey pig iron which contain very small amounts of carbon and other impurities but 1.2-3% slag

c.   Steel which contain from 0.08 to 0.8% carbon

d.   Hard steel which contain 0.8 to 1.5% carbon

e.   Alloy or special steels which besides carbon contain one or more metals such as Ni, Cr, W, V, Mo, Mn.

### 3.4.2.   Raw materials

The main raw materials for the manufacture of iron and steel are iron ore and limes-tone or dolomite as flux. Coking coal is used as fuel. The fuel serves two purposes: to heat the furnace and to produce CO which acts as the reducing agent. To make special steels other materials such as nickel, chromium, cobalt are added.

Iron ore deposits are found in India, China, Brazil, Canada, Germany and United States of America. The ores include red haematite ($Fe_2O_3$), the less inferior brown hydrated haematite also known as limonite

($2Fe_2O_3.3H_2O$), the magnetic magnetite ($Fe_3O_4$) which is black in colour and pyrites ($FeS_2$). The haematite is easily reduced.

Magnetite contains about 72% pure metal and it is reduced with some difficulty.

### 3.4.3.  Removal of impurities in iron ore

The presence of impurities in the iron ore not only reduce the iron content in the ore but also increase production costs especially with regard to consumption of flux and fuel.

If limonite is used, it is first dried before use. When the ore contains large amounts of impurities, appropriate ore dressing operations are carried out on it. When the ore is obtained in small particles, it is sintered into lumps.

The main impurities in iron ore are silica and alumina. Silica and alumina in the presence of limestone makes the ore self-fusing with less production costs. At high temperatures of the blast furnace, the flux reacts with alumina and silica to form a complex of calcium-magnesium aluminium silicate known as slag.

Sulphur and phosphorus are also found in iron ores as impurities in the form of sulphides (FeS), sulphates ($CaSO_4$) and phosphates ($Ca_3(PO_4)_2$ or $Fe_3(PO_4)_2$). Both sulphur and phosphorus, which can also come from the fuel used, are not desired in iron and steel manufacture. Normally steel should not contain more than 0.05% sulphur and 0.05% phosphorus. Sulphur can be removed in the blast furnace slag. Phosphorus cannot be removed in the slag but passes through to the pig iron where it is combined with steel in the convertor. As a result, the ores are sometimes classified as acid or basic ores according to the amount of phosphorus present. Acid ores contain less than 0.05% phosphorus while basic ores has more than 0.05%.

A small amount of manganese is generally present in iron ores. Manganese is advantageous for steel production because it reduces the effect of sulphur by forming manganese sulphide (MnS). Sometimes, if manganese is absent from the ores, it is added.

### 3.4.4.    Fuel

Coke is the fuel used to melt the ore and also to reduce the iron ore to metallic iron. Coke is produced at the bottom of the blast furnace by carbonization of coal i.e. bur-ning of coal in the absence of oxygen to remove volatile matter. Good quality coke has about 80% carbon and 20% ash. It is hard to prevent the formation of CO and its high porosity provides large surface area for the chemical reactions. It is consumed at the rate of one ton per ton of pig iron.

### 3.4.5.    Manufacture of Pig Iron

Pig iron is a direct product of smelting iron ore with fluxes and fuel in a tall blast furnace. The oxygen is introduced at the top of the furnace, blown or blasted through bronze or copper nozzles over the furnace materials in a number of symmetrically placed tubes, called tuyeres. The air blast is preheated to a temperature of about $700^0C$ and pressure of $2.5kg/cm^2$ using the hot exhaust gases leaving the furnace at the top. Preheating greatly increases the economy of steel production.

The molten iron and slag collect at the bottom of the furnace while the gases escape from the top. The slag layer floats over the heavier iron and is periodically collected as dross and stored as waste material that can be used for cement manufacture or for making floor tiles

The pig iron is tapped and is either used to produce cast iron, stored in pigs of sand bags or is taken for steel production. To make cast iron, the molten metal is poured into moulds of desired size and shape. The metal gets cooled and solidifies taking the desired shape.

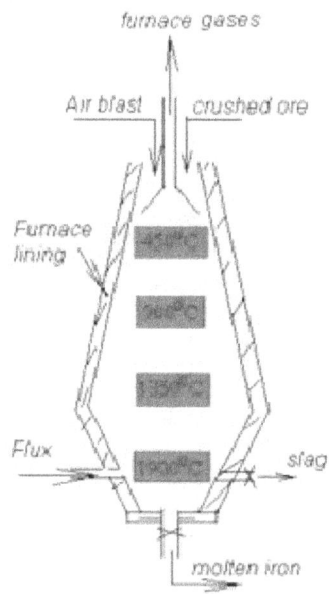

**Fig 3.2.Schematic diagram of a blast furnace showing the temperatures at relative heights.**

## 3.4.6.  Reactions of the blast furnace

The temperature of the blast furnace progressively rises up from top to bottom. The following reactions take place at different zones:

Iron ore reduction:

$$Fe_2O_3 + CO \longleftrightarrow CO_2 + 2 Fe_3O_4$$

$$Fe_3O_4 + CO \longleftrightarrow 3FeO + CO_2$$

$$3FeO + 3CO \longleftrightarrow 3Fe + 3CO_2$$

Fuel reactions

$$C + O_2 \longrightarrow CO_2$$

$$CO_2 + C \longleftrightarrow 2CO$$

Slag formation reactions

$$CaCO_3 \longleftrightarrow CaO + CO_2$$

$$CaO + SiO_2 \longrightarrow CaSiO_3$$

$$2Fe + SiO_2 \longrightarrow 2FeO + Si$$

$$2Mn + SiO_2 \longrightarrow MnO + Si$$

$$MnO_2 + 2C \longrightarrow Mn + 2CO$$

$$FeS + CaO + C \longrightarrow CaS + Fe + CO$$

$$FeS + Mn \longrightarrow Fe + MnS$$

$$Ca_3(PO_4)_2 + 3SiO_2 + 5C \longrightarrow 3CaSiO_3 + 2P + CO.$$

Most of the sulphur passes into the slag as CaS and MnS and only a small portion remains in the metal as FeS and MnS.

## 3.5. Extractive Metallurgy Of Aluminium

Aluminium is the most abundant metal in earth and is commercially extracted from bauxite ores in which it occurs as hydrated aluminium oxide.

*Extraction* of *aluminium from bauxite is carried out in three stages:*

- **Ore dressing**: cleaning ore by means of separation of the metal containingmineral from the waste (gangue).

- **Chemical treatment of bauxite** for converting the hydrated aluminium oxideto pure aluminum oxide.

- **Reduction of aluminium from aluminium oxide** by the electrolytic pro-cess.

Ore dressing may involve washing the ore, size classification and leaching.

### 3.5.1. Chemical treatment of bauxite

At this stage bauxite is crushed and ground to the correct particle size for efficient extraction of the alumina through digestion with hot sodium hydroxide solution which dissolves the aluminium hydroxide, forming a solution of sodium aluminate.

$$2NaOH + Al_2O_3 \longrightarrow Na_2Al_2O_3 + H_2O$$

The residual impurities (oxides of silicon, iron, titanium and aluminium i.e. $SiO_2$, $Fe_2O_3$, $TiO_2$, $Al_2O_3$). These insoluble impurities are called "red mud" which together with fine solid impurities, are separated from the sodium aluminate solution by washing and thickening. The solution is then seeded with aluminium hydroxide from a previous batch in precipitator tanks, where aluminium hydroxide precipitates from the solution.

$$Na_2Al_2O_3 + 4H_2O \longleftrightarrow 2Al(OH)_3 + 2NaOH$$

The aluminium hydroxide after separation from the sodium hydroxide is converted into pure aluminium oxide by heating to 1800F (1000°C) in rotary kilns or fluidized bed calciners.

$$2Al(OH)_3 \longrightarrow Al_2O_3 + 3H_2O$$

### 3.5.2. Reduction of aluminum from aluminium oxide

Primary aluminium is produced by the electrolytic reduction of the aluminium oxide. As aluminium oxide is a very poor electricity conductor, its electrolysis is carried out in a bath of molten cryolite (mineral, containing sodium aluminium fluoride

– $Na_3AlF_6$) as shown in the schematic diagram below.

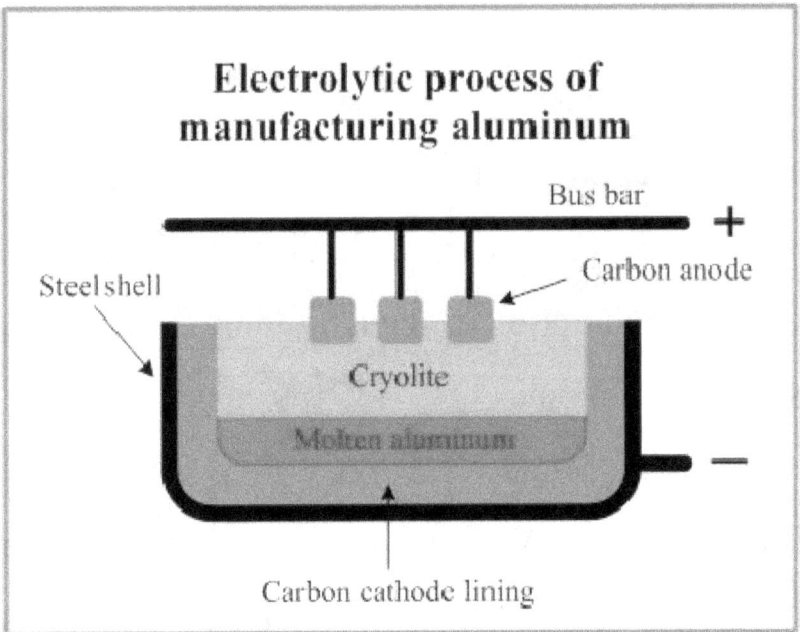

**Fig 3.3** Electrolytic process for aluminium manufacture

This technology is called **Hall-Heroult process.** The electrolytic cell for aluminum production consists of a pot with carbon lining The carbon lining is contained in a steel shell with a thermal insulation of alumina or insulating brick.. This carbon lining serves as the negative electrode (cathode). Prebaked carbon anodes are connected and suspended from the current conductor (bus bar). The anodes are immersed into the bath of molten cryolite at 915 to 950 $^{\circ}$C. The aluminum oxide is added to the cryolite and dissolved in it. When electric current passes between the anodes and the cathode through the cryolite, aluminium oxide decomposes to metallic aluminium deposited at the cathode and oxygen is liberated at the anode. Oxygen from the alu-mina dissolved in the bath combines with the bottom surface of the carbon anode to form carbon dioxide.

Control of alumina concentration in the cells is accomplished by a slight underfeeding. When the alumina reaches a critical level, the cell goes on anode effects caused by a limiting rate of diffusion of alumina to the anode surfaces. The cell voltage then rises and some fluorocarbons are generated. A light bulb connected across the cell lights up with increased cell voltage as a signal for the operators to feed the cell with alumina and kill the anode effect. Cells now run a day or longer between anode effects. The ratio of sodium fluoride to aluminium fluoride in the cryolite bath changes over time and corrective additions are added based on laboratory analyses.

In operation, cryolite freezes on the sidewalls of the cells forming a "ledge" which protects the sidelining from severe attack by aluminium and molten cryolite. Cryolite also freezes over the top of the bath and forms a "crust" to support a top layer of alumina thermal insulation. Alumina is fed to the bath through holes punched in the crust. The carbon dioxide exits through holes in the crust and is collected under the hoods. The carbon dioxide and air leaking in is now ducted to dry scrubbers which remove fluorides from the gas stream. Fresh alumina contacting the gases removes the hydrogen fluoride and evaporated fluoride particulate. This alumina, fed to the cells, returns fluoride to the cells. The hydrogen fluoride comes from residual hy-drocarbons in the anodes and trace water in the alumina and air humidity reacting with the fluoride bath.

The anodes are consumed in the process through the reaction of carbon and oxygen. Replacements are added at individual locations on a regular schedule. The anode butts are sent back to the anode plant to be ground and mixed into new anode paste to be pressed and baked.

The molten aluminium is periodically tapped under vacuum from the furnace into a crucible and cast into ingots.

## 3.6. Extractive Metallurgy Of Copper

Copper is mostly extracted from ores containing copper sulphides, copper oxides or copper carbonates. Copper ores are generally poor and contain between 1.5 and 5% copper. Therefore, commercial extraction of copper involves several dressing operations before the smelting stage.

The extraction of copper from its sulphide ores is done by eliminating the gangue, iron, sulphur and minor impurities by the following steps yielding the shown % copper after each step:

|  | % copper |
|---|---|
| • **Concentration** | 15-25 |
| • **Roasting** | 30-45 |
| • **Smelting** | - |
| • **Matte conversion** | 98 |
| • **Fire refining** | 99.5 |
| • **Electrolytic conversion** | 99.9 |

### 3.6.1.    Concentrating

The purpose of concentration step is to separate the copper mineral from the gangue.

The ore is first crushed and finely ground. It is made into a slurry with water and then fed into a froth flotation cell. The ore particles are lifted up by air bubbles while the gangue remain in the cell. The froth containing the ore is thickened and filtered.The pulp is dried to about

6% moisture.

### 3.6.2. Roasting

The objective of roasting is to remove excess sulphur. Thus, if the ore does not contain excess sulphur, roasting may be omitted and the ore directly smelted. Roasting is carried out in a multiple hearth furnace or in a fluidized bed.

The dry pulp is fed into the roaster at 600 to 700 oC. The burning of the sulphide ores supplies the heat to maintain the temperature at which roasting takes place. The reactions at the roaster are as follows.

$$2As_2S_3 + 9O_2 \rightarrow 2As_2O_3 + 6SO_2 \quad 2Sb_2S_3 + 9O_2 \rightarrow 2Sb_2O_3 + 6SO_2$$

$$Fe_2S_3 \rightarrow 2FeS + SS + O_2 \quad SO_2$$

$$2FeS + 3O_2 \rightarrow 2SO_2 + 2FeO \quad 4FeO + O_2 \rightarrow 2Fe_2O_3$$

The arsenic and antimony are volatiles and and leave with sulphur dioxide.

### 3.6.3. Matte smelting

At this stage the concentrate is smelted in a furnace to produce a mixture of copper and iron, called matte.

Smelting is carried out at about $1350^{\circ}C$. The rasted ore is in powder form and cannot therefore be smelted conveniently in a blast furnace. It is done in a long reverbera-tory furnace heated by coal dust. The following are the reactions that take place in the furnace:

$$2FeS + 3O_2 \rightarrow 2FeO + 2SO_2 FeO + SiO_2 \rightarrow FeSiO_3$$

$$Cu_2O + FeS \rightarrow Cu_2S + FeO$$

$$CuO + FeS \rightarrow CuS + FeO2CuS \rightarrow Cu_2S + S$$

$$3Fe_2O_3 + FeS \rightarrow 7FeO + SO_2$$

$Cu_2S$ and part of the FeS form the matter. The silicates and oxides of iron are slagged. The gangue is absorbed by the slag and removed.

### 3.6.4. Blister copper production

The object of the converter operation is to convert matte into molten blister copper containing 96 to 98% copper and remove the iron rich slag

The operation is carried out in two stages each of which has a distinctive flame colour. In the first stage, air is blown through the matte untill all the slag is formed i.e. the total elimination of the FeS. Silica is used to react with the oxide. The slag is removed by tilting the converter. Air is again blown through the matte and the $Cu2S$ is converted to Cu. The following are the reactions in the converter:

Slag formation stage: $2FeS + 3O_2 \rightarrow 2FeO + 2SO_2$

$2FeO + 2SiO_2 \rightarrow 2FeSiO_3$

Blister copper formation $2Cu_2S + 3O_2 \rightarrow 2Cu_2O + 2SO_2$

$2Cu_2O + Cu_2S \rightarrow 6Cu + SO_2$

### 3.6.5. Fire Refining

The blister copper is fed into a furnace where some of the Cu is oxidised into $Cu_2O$ which dissolves in the molten copper. The oxide rapidly oxidizes the impurities. $SO_2$ passes out while other impurities form dross on the surface. The dross is frequently skimmed off to expose fresh surface for oxidation. A pole of green wood is then thrust in and hydrogen from the wood reduces the excess oxygen. Poling is continued until proper surface characteristics of the cooled samples are obtained. The product is called tough pitch. It has good electrical conductivity. It is cast into slabs.

## 3.6.6. Electrolytic refining

Tough pitch copper is not fit for gas-welding until it is deoxidized further. It is made into impure copper anodes which are immersed in a 5 to 10% sulfuric acid bath contaning copper sulphate. Pure copper foil serves as the cathode where copper deposits.

Cathodes produced as a result of the electrolytic refining process contain 99.9% of copper which is used for manufacturing copper and copper alloys products.

## Unit Objectives

At the end of this unit you should be able to:

a.   Describe the various stages mineral ores go through in a typical ore dressing process.

b.   Write equations to describe calcination and roasting

c.   Explain what happens during smelting

d.   Describe the extractive metallurgy of iron

e.   Describe the extractive metallurgy of copper

f.   Describe the extractive metallurgy of aluminium

## Summary of Chapter

In this chapter, we shall study how metals are extracted from mineral ores in which they exist with other materials of less value. Generally, ores are first taken through size reduction, sorting and agglomeration to transform them into a form that can be taken through extraction processes including calcining, roasting, smelting and refining. Extractive metallurgy of iron, aluminium and copper respectively are then presented.

## References

Das R.K. (1988) Industrial Chemistry: Metallurgy, Kalyani Publishers, New Delhi.

## List of relevant resources

- Computer with internet facility to access links and relevant copywrite free resources

- CD-Rom accompanying this module for compulsory reading and demonstrations

- Multimedia resources like video, VCD, and CD players

## List of relevant useful links

http://www.mine-engineer.com

http://electrochem.cwru.ed/encycl

The first site has useful informationon on various unit operations used in the chemical industry. Photographs and other illustrations are given. The second site has information on aluminium production.

# CHAPTER – 4

## Chlor-alkali, Ammonia, Sulphuric Acid, Fertilizer, Cement

☞CONTENTS

## 4.1 Manufacture of Sodium Hydroxide and Chlorine by the Chlor-alkali process

### 4.1.1. Introduction

Before the electrolytic method of making sodium hydroxide and chlorine (chlor-alkali process) became widely used, sodium hydroxide was made from soda ash by the **lime-soda** process. The soda ash as aqueous $Na_2CO_3$ is reacted with slaked lime $(Ca(OH)_2)$ according to the following equation:

$$Na_2CO_{3(aq)} + Ca(OH)_{2(s)} = 2NaOH_{(aq)} + CaCO_{3(s)}$$

The chlor-alkali process has gradually replaced the lime-soda process.

### 4.1.2. Raw Materials

The chlor-alkali industry uses rock salt, a natural deposit of sodium chloride. The aqueous sodium chloride solution, referred to as brine contains $Na^+$, $Cl^-$, $H^+$ and $OH^-$ ions. Electrolysis of this solution produces simultaneously chlorine, sodium hydroxide and hydrogen in the ratio of 1:1.13:0.028

### 4.1.3. Chlor-alkali process

The term chlor-alkali refers to the two chemicals (chlorine and an alkali) which are simultaneously produced as a result of the electrolysis of brine. The most common chlor-alkali chemicals are chlorine and sodium hydroxide (caustic soda) but can include potassium hydroxide when a potassium brine is used.

Brine electrolysis produces chlorine at the anode and hydrogen along with the alkali hydroxide at the cathode. The two products are removed in separate streams.

The overall chemical reaction of electrolysis of sodium chloride is: $2NaCl_{(aq)} + 2H_2O_{(l)} = 2NaOH_{(aq)} + Cl_{2(g)} + H_{2(g)}$

If chlorine is not separated from the sodium hydroxide, side reactions such as the following would occur:

$2NaOH_{(aq)} + Cl_{2(g)} = NaOCl_{(aq)} + NaCl_{(aq)} + H_2O_{(l)}$

$H_{2(g)} + Cl_{(g)} = 2HCl_{(g)}$

Three processes are in use: the diaphragm-cell process, the membrane-cell process, and the mercury-cell process. These are shown in the figures below.

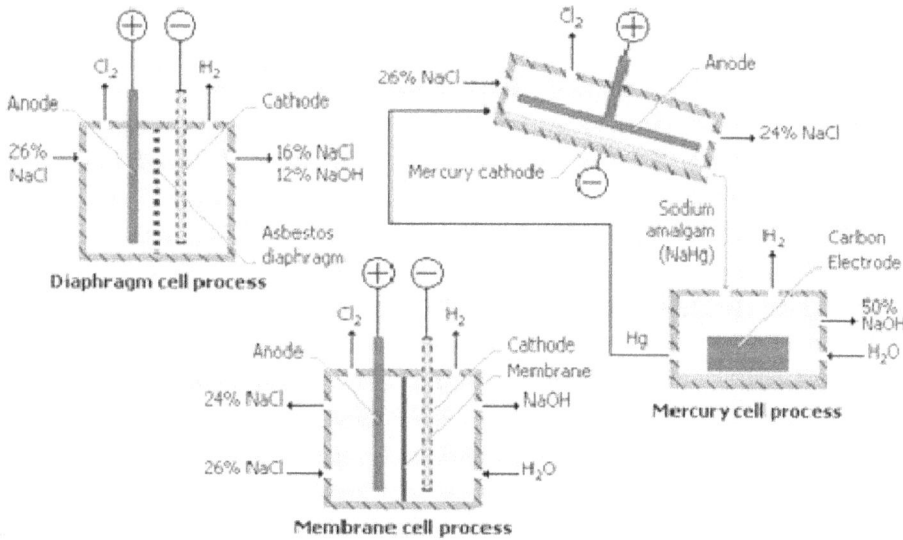

**Fig 4.1.**    Chlor-alkali cells

## 4.1.4.   Mercury Cells

In the mercury-cell process, a flowing pool of mercury at the bottom of the electro-lytic cell serves as the cathode. **The anodes are graphite or modified titanium.** When an electric current passes through the brine, chlorine is produced at the anode and sodium dissolves in the mercury, forming an amalgam of sodium and mercury. The amalgam is then poured into a separate vessel, where it decomposes into sodium and mercury.

Cell reactions

At the anode

$$2Cl^-_{(aq)} \rightarrow Cl_{2(g)} + 2e^-$$

At cathode

$$2Na^+_{(aq)} + 2e^- \rightarrow 2Na/Hg$$

Overall: $2NaCl_{(aq)} \rightarrow 2NaHg + Cl_{2(g)}$

The amalgam (a mercury-sodium alloy) is taken to a different vessel for decomposition according to the equation:

$2Na \cdot Hg + 2H_2O = 2NaOH + H_2 + Hg$

Initially mercury cells seemed to dominate the field because it produced high quality products. With exactly the required amount of water added, a 50%NaOH solution is formed which does not greatly require much evaporation for concentration.

However, mercury cells use much more electrical energy than diaphragm and membrane cells. Also, small quantities of mercury discharges into nearby streams. These discharges were found to be sources of the carcinogenic methyl mercury. This has led to prohibition and gradual replacement of the mercury cells by diaphragm cells.

## 4.1.5. Diaphragm Cells

These cells contain a diaphragm, usually made of asbestos fibres that separate the anode from the cathode. The diaphragm also allows ions to pass by electrical migration but limit the diffusion of products. The anodes are made of graphite and the cathodes of cast iron. When an electric current passes through the brine, the chlorineions and sodium ions move to the electrodes. Chlorine gas is produced at the anode, and sodium ions at the cathode react with the water, forming caustic soda. Some salt remains in the solution with the caustic soda and can be removed at a later stage.

Advantages and disadvantages

i.   The electrodes can be placed close together and this permits compactness of diaphragm cells of low electrical resistance.

ii.  They easily become congested with use (shown by high voltage drop) and therefore must be replaced regularly.

iii. Diaphragms permit flow of brine from anode to cathode and this reduces greatly on side reactions like formation of sodium hypochlorite.

iv. Diaphragm cells with metal cathodes like titanium coated with rare earth oxides, rarely develop congested diaphragms and do not require regular replacement. This reduces on operating costs.

v. The release of asbestos to the environment is a major anticipated problem of diaphragm cells. Diaphragms made of corrosion resistant plastics are a proposed solution to asbestos pollution.

vi. A major advantage is that diaphragm cells can operate on dilute 20% reaso-nably impure brine. Such brines produce dilute sodium hydroxide (about 15%) contaminated with sodium chloride. Concentrations to about 50% are achieved by use of multi-effect evaporators. Studies have shown that nearly 2,600kg of water must be evaporated to obtain 1 ton of 50% caustic soda.

Diaphragm cell technology is still in use, but the use of asbestos in the cells is causing new plants to turn to the newer ion-exchange or membrane technology.

## 4.1.6. Membrane Cells

Improved designs of membrane cells and cheaper purification have increased the economics of the new membrane process. Fig 4.2 represents a condensed overview of a membrane cell process.

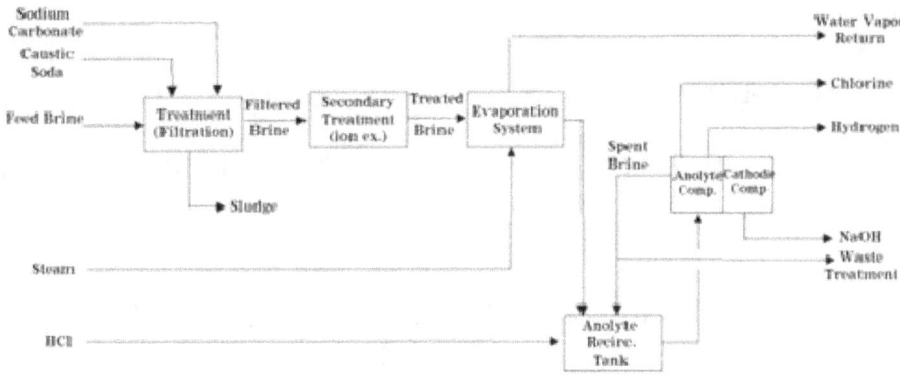

**Fig 4.2**   Membrane chlor-alkali process

In this process, rather than using a diaphragm, semi permeable membrane made of plastic sheets is used to separate the anode and cathode compartments. The plastic sheets are porous and chemically active to allow sodium ions to pass through but reject hydroxyl ions. An example of membrane material is perfluorosulphonic acid polymer.

The membranes exclude OH⁻ and Cl⁻ ions from the anode chamber thereby making the product free from the salt contamination experienced in the diaphragm cell. The membrane cells use more concentrated brine solutions and produce purer and more concentrated sodium hydroxide, which require less evaporation during concentration.

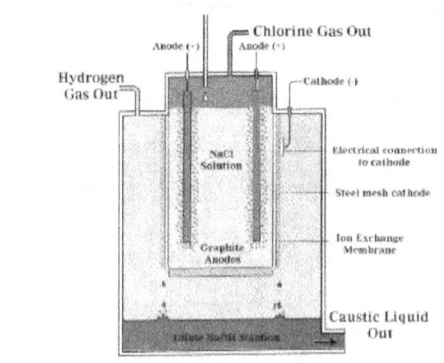

**Fig 4.3**      Chlor-alkali membrane cell

Brine is pumped into the anode compartment, and only sodium ions pass into the cathode compartment, which contains pure water. Thus, the caustic soda producedhas very little salt contamination.

## 4.2. Manufacture of Ammonia

### 4.2.1.    Introduction.

Ammonia is one of the most highly produced inorganic chemicals in the world be-cause of its widespread application. Synthetic ammonia is produced from the reaction between nitrogen and hydrogen. Before synthetic nitrogen fixation was discovered, manures, ammonium sulfate (a by-product from the coking of coal), Chilean saltpetre, and later, ammonia recovered from coke manufacture were some of the important sources of fixed nitrogen. During the first decade of the twentieth century, the world-wide demand for nitrogen-based fertilizers far exceeded the existent supply.

### 4.2.2.    Uses of ammonia

Ammonia is the basis from which virtually all nitrogen-containing products are derived.

The main uses of ammonia include the manufacture of:

- Fertilizers ((ammonium sulfate, diammonium phosphate, urea)
- Nitric acid
- Explosives
- Fibres, synthetic rubber, plastics such as nylon and other polyamides
- Refrigeration for making ice, large scale refrigeration plants, air-conditioning units in buildings and plants

- Pharmaceuticals (sulfonamide, vitamins, etc.)
- Pulp and paper
- Extractive metallurgy
- Cleaning solutions

### 4.1.3.    Raw Materials

The raw materials used to manufacture ammonia are air, water and, hydrocarbons. Coal can also be used in place of hydrocarbons but the process is complex and ex-pensive.

### 4.2.4.    Nitrogen fixation

For a long time, commercial development of nitrogen fixation ammonia process had proved elusive. Old methods used to produce ammonia included dry distillation of nitrogenous vegetable and animal waste products. Here, nitrous acid and nitrites were reduced with hydrogen according to the following equation:

$$N_2O + 4H_2 = 2NH_3 + H_2O$$

Ammonia was also produced by the decomposition of ammonium salts using alkaline hydroxides such as quicklime as shown in the following equation.

$$2NH_4Cl + 2CaCl_2 = CaCl_2 + Ca(OH)_2 + 2NH_3$$

Haber invented a large-scale catalytic synthesis of ammonia from elemental hydro-gen and nitrogen gas, reactants which are abundant and inexpensive. By using high temperature (around $500^{\circ}C$), high pressure (approximately 150-200 atm), and an iron catalyst, Haber could force relatively unreactive gaseous nitrogen and hydrogen to combine into ammonia.

The collaborative efforts of Haber and Carl Bosch made the commercial high-pressure synthesis of ammonia possible by 1913. This energy-intensive process has undergone considerable modification in recent years.

### 4.2.5. Chemical Reaction and Equilibrium

Ammonia synthesis from nitrogen and hydrogen is an exothermic reversible reaction and can be described by the following overall reaction.

$$1/2N_2 + 3/2H_2 \xrightarrow[\longleftrightarrow]{500\,^{\circ}C,\ Fe} NH_3 \qquad H = 45.7 kJ/mol$$

The reaction is accompanied by decrease in volume and by Le Chatelier's principle, increasing the pressure causes the equilibrium to shift to the right resulting in a higher yield of ammonia. Since the reaction is exothermic, decreasing the temperature also causes the equilibrium position to move to the right again resulting in a higher yield of ammonia. We can conclude then that ammonia synthesis as per equation (1) is an equilibrium reaction that is favoured by low temperature and high pressure. Thermodynamics gives us equilibrium conditions of the reaction but does not give us any idea about the rate of reaction. The reaction does not proceed at ambient temperature because nitrogen requires a lot of energy to dissociate. In the gas phase this dissociation occurs only at around 3000°C. Even the hydrogen molecule, which has a weaker molecular bond, only dissociates markedly at temperatures above 1000°C.

### 4.2.6. Catalyst

Since the ammonia synthesis reaction cannot be moved to the right at low temperature, this calls for temperature increase, which unfortunately drives the reverse reaction. This is where the role of the iron catalyst comes in. The reaction when carried out at high pressures and temperatures occurs with large yields when iron catalysts are present. The hydrogen and nitrogen molecules lose their translational degrees of freedom when bound to the catalyst surface. This reduces the activation energy for the release of atomic nitrogen dramatically, and thus makes the forward reaction go faster at lower temperatures. The use of lower temperature reaction conditions means there is limited reverse reaction. But we still need reasonably high temperatures (250-400°C) even with the use of a catalyst which essentially accelerates the reac-tion sufficiently so that we can obtain ammonia at conditions where the equilibrium conversion is large enough to be useful.

### Iron catalyst

The real problem has been to find a suitable catalyst so that the maximum amount of product is obtained with minimum volume of the catalyst in the shortest time possible. The catalysts, which Haber initially employed, were catalytically active pure metals, which were too expensive. They also lost catalytic activity after a short time due to poisoning. Iron was one of the metals that showed high activity. However, pure iron has a regular shape with low porosity which is a disadvantage for a catalyst. The iron can be made irregular by mixing and fusing iron oxide with other oxides combinations. This is referred to as **structural promotion**. A common iron catalyst promoter is a mixture of $Al_2O_3$ and $K_2O$. Its optimal performance requires reaction temperatures around 400°C and pressures from 150-300 atmospheres. Ammonia synthesis reaction is carried out commercially at 200-500 atm and 450 to 600°C.

## 4.2.8. Modern Method of Manufacturing Ammonia

The manufacturing process consists of six stages namely: manufacture of reactant gases, purification, compression, catalytic reaction, recovery of ammonia formed and recirculation and ammonia removal as shown in the flow diagram Fig. 4.4.

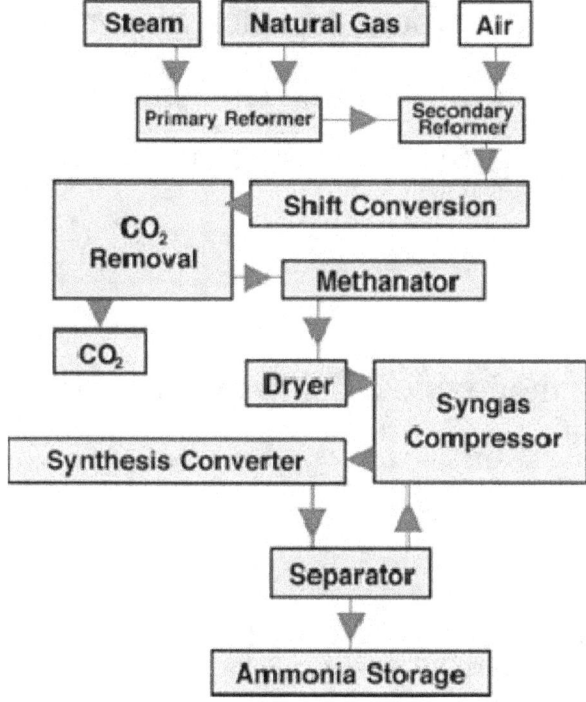

**Fig. 4.4** Block diagram of an ammonia plant

Hydrogen is obtained by conversion of hydrocarbons such as methane, propane, butane or naphtha into gaseous hydrogen.

## 4.2.8.1. Desulphurization

Hydrocarbon feedstocks contain sulphur in the form of $H_2S$, COS, $CS_2$ and mercaptans. The catalyst used in the reforming reaction is deactivated (poisoned) by sulphur. The problem is solved by catalytic hydrogenation of the sulphur compounds as shown in the following equation:

$$H_2 + RSH = RH + H_2S_{(g)}$$

The gaseous hydrogen sulphide is then removed by passing it through a bed of zinc oxide where it is converted to solid zinc sulphide:

$$H_2S + ZnO = ZnS + H_2O$$

## 4.2.8.2. Primary (Steam) Reforming.

Reforming is the process of converting natural gas or naptha ($C_nH_{2n+2}$) into hydro-gen, carbon monoxide and carbon dioxide. Steam and natural gas are combined at a three-to-one ratio. This mixture is preheated and passed through catalyst-filled tubes in the primary reformer.

Catalytic steam reforming of the sulphur-free feedstock produces synthesis gas (hydrogen and carbon monoxide). Using methane as an example:

$$CH_4 \underset{}{\overset{Ni, \ 15\text{-}20 \ atm, \ 1000\text{-}1100^{oC}}{\rightleftarrows}} CO + 3H_2$$

The reaction is endothermic. It is operated at $1000\text{-}1100^{oC}$. It is not favoured by high pressures, but to reduce volumetric flow rate at high temperature, the steam reforming reaction is carried out at high pressures of 15 to 20 atm.

## 4.2.8.3. Secondary reformer

From the primary reformer, the mixture flows to the secondary reformer. Air is fed into the reformer to completely convert methane to CO in the following endothermic reaction.

$$CH_4 + Air \xrightleftharpoons[Ni, 15\text{-}20\, atm,\, 1000\text{-}1100\,^{oC}]{} CO + H_2O + N_2$$

The nitrogen and hydrogen coming out of the secondary reformer are in the ratio of 3:1. This mixture is known as the synthesis gas.

### 4.2.8.4. Shift Conversion.

The carbon monoxide is converted to carbon dioxide with the assistance of catalyst beds at different temperatures.

$$CO + H_2O = CO_2 + H_2$$

This water-gas shift reaction is favorable for producing carbon dioxide which is used as a raw material for urea production. At the same time more hydrogen is produced.

### 4.2.8.5. Purification.

The carbon dioxide is removed either by scrubbing with water, aqueous monoetha-nolamine solution or hot potassium carbonate solution.

CO is an irreversible poison for the catalyst used in the synthesis reaction, hence the need for its removal The synthesis gas is passed over another catalyst bed in the methanator, where remaining trace amounts of carbon monoxide and dioxide are converted back to methane using hydrogen.

$$CO + 3H_2 = CII_4 + H_2O$$

$$CO_2 + 4H_2 = CH_4 + 2H_2O$$

$$O_2 + 2H_2 \longrightarrow 2H_2O$$

Note that the first equation is the opposite of the reformer reaction.

### 4.2.8.6. Ammonia Converter.

After leaving the compressor, the gaseous mixture goes through catalyst beds in the synthesis converter where ammonia is produced with a three-to-one hydrogen-to-nitrogen stoichiometric ratio. However, not all the hydrogen and nitrogen are converted to ammonia. The unconverted hydrogen and nitrogen are separated from the ammonia in the separator and recycled back to the synthesis gas compressor and to the converter with fresh feed. Because the air contains argon which does not participate in the main reactions, purging it minimizes its build up in the recycle loop.

### 4.2.8.7. Ammonia Separation

The removal of product ammonia is accomplished via mechanical refrigeration or absorption/distillation. The choice is made by examining the fixed and operating costs. Typically, refrigeration is more economical at synthesis pressures of 100 atm or greater. At lower pressures, absorption/distillation is usually favoured.

### 4.2.8.8. Ammonia Storage

Ammonia is stored in tanks as a refrigerated liquid. Some ammonia is used directly as a fertilizer. Most ammonia is converted in downstream processes to urea (46% nitrogen) or ammonium nitrate (34% nitrogen) for use as fertilizer.

### 4.2.9. Some environmental impacts of ammonia production

Ammonia is toxic, irritant and corrosive to metal alloys (e.g. copper alloys). In refrigeration, its replacement by the non-toxic chlorofluorocarbon (CFCs) has contributed to global warming. In large industrial processes such as bulk ice making, and food processing and preservation, ammonia is still being used as a refrigerant.

The toxicity of ammonia solution is usually not harmful to human beings because it is easily excreted in urine. However, ammonia even in dilute concentrations is toxic to aquatic animals because they do not have the mechanisms to eliminate it from their bodies by excretion.

## 4.3. Sulphuric Acid Manufacture

### 4.3.1. Introduction

During the 19th century, the German chemist Baron Justus von Liebig discovered that sulphuric acid, when added to the soil, increased the amount of soil phosphorus available to plants. This discovery gave rise to an increase in the commercial production of sulphuric acid and led to improved methods of manufacture.

### 4.3.2. Uses of sulphuric acid

Sulphuric acid is the most widely used chemical. The largest single use of sulphuric acid is for making phosphate and ammonium sulphate fertilizers. Other uses include production of phosphoric acid, trisodium phosphates for detergent making. Sulphuric acid is also used in large quantities in iron and steel making as a pickling agent to remove oxidation, rust and scale from from the metals. It is an oxidizing and dehy-drating agent. Its dehydrating action is vital in absorbing water formed in chemical conversions such as nitration, sulphonation, and esterification. It vigorously removeswater from, and therefore chars, wood, cotton, sugar, and paper. As a strong oxidizing agent it is capable of dissolving such relatively unreactive metals as copper, mercury, and lead to make compounds of these metals.

It is used in the manufacture of aluminium sulphate for application in paper pulpproduction and in water treatment. It is also used as an electrolyte in lead acid batteries found in cars.

Various concentrations of sulphuric acid are available depending on the application purpose. These include:

- 10% dilute acid for laboratory use, pH = 1
- 33.3% for lead acid batteries, pH = 0.5
- 62.2% for chamber and fertilizer manufacture, pH = 0.4
- 77.7% tower or Glover acid pH = 0.25
- 93.2% Oil of Vitriol
- 98% conc acid, pH = 0.1
- 100% $H_2SO_4$
- 20% oleum (104.5% $H_2SO_4$)

### 4.3.3. Raw Materials

Raw materials for sulphuric acid are those that produce sulphur dioxide when reacted with oxygen. The commonly used raw materialds are:

- Elemental sulphur
- Sulphides such as pyrites
- Hydrogen sulphide from petroleum refineries

### 4.3.4. Manufacturing process

Two processes, the lead-chamber and contact processes, are used for the production of sulphuric acid. In their initial steps, both processes require the use of sulphur dioxide.

## 4.3.4.1. The Lead-chamber process

This process employs as reaction vessels large lead-sheathed brick towers. In these towers, sulphur-dioxide gas, air, steam, and oxides of nitrogen react to yield sulphuric acid as fine droplets that fall to the bottom of the chamber. Almost all the nitrogen oxides are recovered from the outflowing gas and are brought back to the chamber to be used again. Sulphuric acid produced in this way is only about 62 to 70 per cent $H_2SO_4$. The rest is water. The chamber process has become obsolete and has been repalced by the **contact** process due to the following reasons:

i.    An increased demand for strong, pure acid and oleum

ii.   Contact process plants are cheaper and more compact

## 4.3.4.2. The Contact Process

The second method of manufacturing sulphuric acid, the contact process, which came into commercial use about 1900, depends on oxidation of sulphur dioxide to sulphur trioxide, $SO_3$, under the accelerating influence of a catalyst.

The first contact plants (before 1920) were built using platinum catalysts. Finely divided platinum, the most effective catalyst, has two disadvantages: it is very expensive, and it is deactivated by certain impurities in ordinary sulphur dioxide. They includecompunds of arsenic, antimony and lead. In the middle of 1920s, vanadium catalysts started being used and have since then replaced platinum. By 1930, the contact process could compete with the chamber process and because it produces high strength acid, it has almost replaced the chamber process.

Since the oxidation of sulphur and sulphur dioxide releases large amounts of energy, major changes in the manufacturing plant design were introduced to utilise this heat energy in the production of steam for generating electrical power. This combination of a chemical plant and electrical generation is known as co-generation.

The flow diagram for sulphuric acid manufacture by the contact process is shown in Fig. 4.5

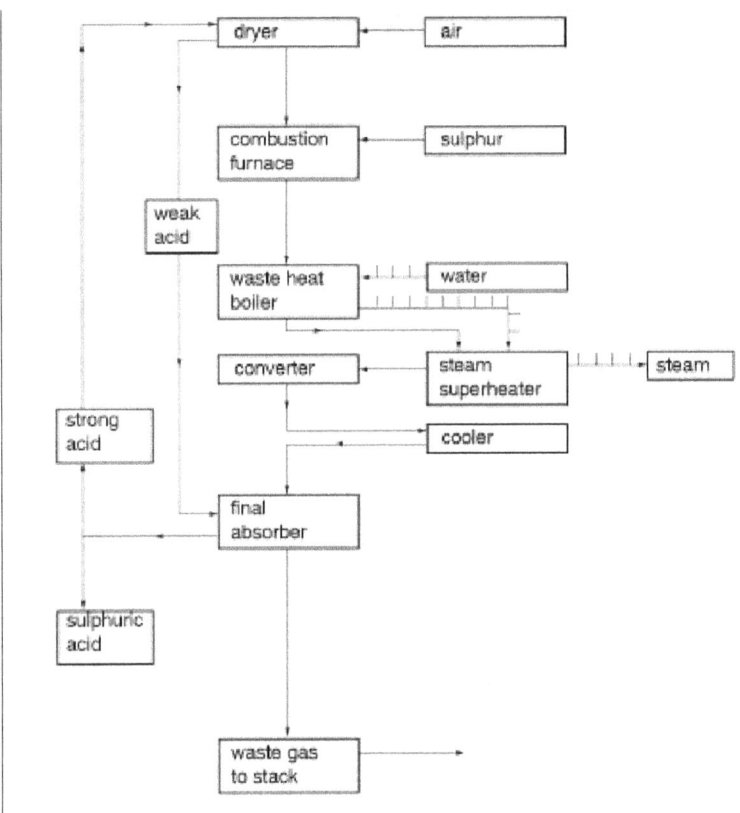

**Fig. 4.5** Block diagram for the manufacture of sulphuric acid by the contact process.

The main steps in the contact plant are:

- production of sulphur dioxide gas

- purifying and cooling the gas

- the gas conversion of $SO_2$ into sulphur trioxide ($SO_3$) by passing it through a converter containing the catalyst

- absorbing the sulphur trioxide in sulphuric acid

## 4.3.4.3. Production of $SO_3$

Sulphur is burned in the sulphur burner to produce sulphur dioxide:

**$S(s)+O_2(g) = SO_2(g)$, $\Delta H$ = -298.3 kJat $25^0C$**

Before combustion, sulphur, is first melted by heating it to 135°C. Combustion is carried out at between 900 and 1800°C. The combustion unit has a process gas cooler. The $SO_2$ content of the combustion gases is generally around 18% by volume and the $O_2$ content is low but higher than 3%. The gases are generally diluted to 9-12% $SO_2$ before entering the conversion process.

*Conversion of $SO_2$ into $SO_3$*

The design and operation of sulphuric acid plants are focused on the following gas phase chemical reaction in the presence of a catalyst:

**$2SO_2+O_2(g) = 2SO_3(g)$, $\Delta H$ = -98.3kJ at $25^0C$**

From thermodynamic and stoichiometric considerations, the following methods are available to maximise the formation of $SO_3$ for the $O_2/SO_2/SO_3$ system.

- heat removal: the formation of $SO_3$ is exothermic, so a decrease of temperature will be favourable

- increased oxygen concentration

- removal of $SO_3$
- raising the system pressure
- catalyst selection to reduce the working temperature
- longer reaction time

This reaction is a reversible reaction and the conditions used are a compromise between equilibrium and rate considerations.

It is necessary to shift the position of the equilibrium as far as possible to the right in order to produce the maximum possible amount of sulphur trioxide in the equilibrium mixture. Even though excess $O_2$ would move the $SO_2$ formation to the right, the 1:1 mixture gives the best possible overall yield of sulphur trioxide. The forward reaction is exothermic and is favoured by low temperature. However, too low a temperature slows the reaction. To get the gases to reach equilibrium within a very short time, a compromise temperature of 400–450°C is used. According to Le Chatelier's principle high pressures favour the forward reaction. However, even at relatively low pressures of 1 to 2 atmospheres, there is a 99.5% conversion of sulphur dioxide into sulphur trioxide. In the absence of a catalyst the reaction is quite slow and is therefore carried out in the presence of a vanadium oxide catalyst which has a long life because it is not easily poisoned. Further more, vanadium catalyst has high conversion efficiency.

Its only disadvantage is that it requires use of low sulphur dioxide concentration which makes plant capital cost to be high.

In summary, optimum conditions for sulphuric acid production in the contact process are:

- A temperature of about $430^0C$

- A pressure of 2 atmospheres

- Vanadium pentoxide catalyst.

A diagram of the converter is shown in Fig. 4.6

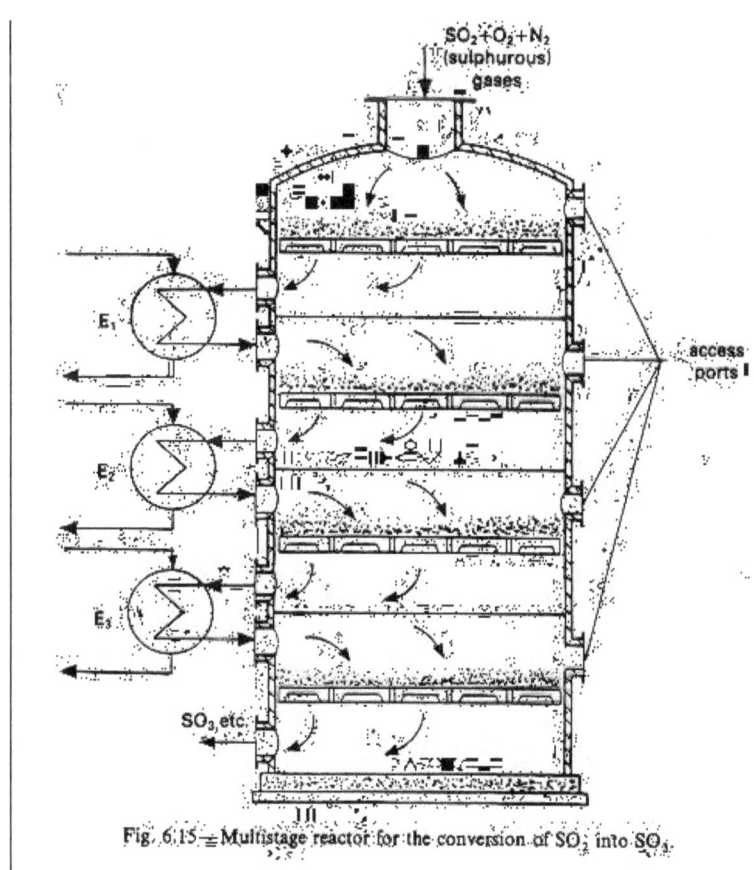

Fig. 6.15 – Multistage reactor for the conversion of SO₂ into SO₃.

**Fig 4.6**    Converter for $SO_2$ into $SO_3$

## 4.3.4.4. Absorption of $SO_3$

Sulphuric acid ($H_2SO_4$) is obtained from the absorption of $SO_3$ into sulphuric acid with a concentration of at least 98%, followed by the adjustment of the strength by the controlled addition of water. $SO_3$, will react with water to form sulphuric acid. However, converting the sulphur trioxide into sulphuric acid cannot be done by simply adding water to the sulphur trioxide. Direct mixing of sulphur trioxide with water by the following reaction is uncontrollable. The exothermic nature of the reaction means it generates a fog or mist of sulphuric acid, which is more difficult to work with than a liquid.

**$SO_3(g) + H_2O(l) = H_2SO_4(l)$, $\Delta H = -130.4kJ$ at $25^0C$**

Instead, the sulphur trioxide is first dissolved in concentrated (98%) sulphuric acid to form a product known as fuming sulphuric acid or oleum.

**$SO_3 + H_2SO_4(l) + = H_2S_2O_7(l)$**

The oleum can then be reacted safely with water to produce concentrated sulphuric acid.

**$H_2S_2O_7(l) + H_2O(l) = 2H_2SO_4(l)$**

## 4.3.4.5. Environmental Issues

Sulphuric acid is a constituent of acid rain, formed by atmospheric oxidation of sulphur dioxide in the presence of water. Sulphur dioxide is released when fuels containing sulphur such as oil and coal are burned. The gas escapes into the atmosphere forming sulphuric acid. Sulphuric acid is also formed naturally by oxidation of sulphide ores.

## 4.4. Manufacture of Fertilizers

## 4.4.1. Introduction

Fertilizer is a substance added to soil to improve the growth and yield of plants. Modern synthetic fertilizers are composed mainly of nitrogen, phosphorus, and potassium compounds with secondary nutrients added.

The process of adding substances to soil to improve its growing capacity was developed in the early days of agriculture. Ancient farmers knew that the first yields on a plot of land were much better than those of subsequent years. This caused them to move to new, uncultivated areas, which again showed the same pattern of reduced yields over time. Eventually it was discovered that plant growth on a plot of land could be improved by spreading animal manure throughout the soil. Over time, fertilizer technology became more refined. New substances that improved the growth of plants were discovered.

## 4.4.2. Uses of fertilizer

The use of synthetic fertilizers has significantly improved the quality and quantity of the food available today, although their long-term use is an environmental subject of debate.

As has been mentioned, fertilizers are typically composed of nitrogen, phosphorus, and potassium as micronutrients. Nitrogen helps make plants green and plays a major role in boosting crop yields. It plays a critical role in protein formation and is a key component of chlorophyll. Plants with adequate nitrogen show healthy vigo-rous growth, strong root development, dark green foliage, increased seed and fruit formation and higher yields.

Plants also need phosphorus, a component of nucleic acids, phospholipids, and several proteins. Phosphorus is also needed to provide the energy to drive metabolic chemical reactions. Without enough phosphorus, plant growth is reduced.

Potassium helps plants grow strong stalks, in the same way that calcium gives people strong bones. It is used in protein synthesis and other key plant processes. Yellowing, spots of dead tissue, and weak stems and roots are all indicative of plants that lack enough potassium.

Besides the above three macronutrients, fertilizers also contain trace elements that improve the growth of plants. Calcium, magnesium, and sulfur are also important materials in plant growth. They are only included in fertilizers in small amounts, however, since most soils naturally contain enough of these components. Other micronutrients include iron, chlorine, copper, manganese, zinc, molybdenum, and boron, which primarily function as cofactors in enzymatic reactions. While they may be present in small amounts, these compounds are no less important to growth, and without them, plants can die. The absence of any one nutrient in the soil can limit plant growth, even when all other plant nutrients are present in adequate amounts.

### 4.4.3. Raw materials

Primary fertilizers include substances derived from nitrogen, phosphorus, and po-tassium. Nitrogen is derived from ammonia, phosphorus from phosphate rock and potassium from potassium chloride, a primary component of potash.

### 4.4.4. The Manufacturing Process

Fully integrated factories have been designed to produce compound fertilizers from primary fertilizers. Depending on the actual composition of the end product, the production process will differ from manufacturer to manufacturer.

Multicomponent fertilizers are compound fertilizers composed of primary nitrogen, phosphorus, and potassium (NPK) fertilizers and secondary nutrients. Generally, each granule of the compound fertilizer contains a uniform ratio of nutrients, or blends. The three numbers on a bag of such a fertilizer are referred to as the "analysis." It is the percentage of nitrogen, phosphate and potash that is available to plants from that bag of fertilizer. N-5-P-10-K-5 or simply 5-10-5 means 5 percent nitrogen, 10 percent phosphate and 5 percent potash. The analysis found on a bag or bulk shipment of fertilizer tells the farmer or consumer the amount of nutrients being supplied.

The balance 80 percent will contain some micronutrients and filler material, which allows for even

application of the nutrients across the fertilized area.

Fig 4.7    A bag of fertilizer showing analysis

### 4.4.4.1. Nitrogen fertilizer

*Ammonia (82-0-0)* is a basic nitrogen fertilizer. Stored as a liquid under pressure orrefrigerated, it becomes a gas when exposed to air and is injected into the soil It is also used as a building block to make other easy to handle nitrogen fertilizer products, including urea, ammonium nitrate, ammonium sulfate and water-based liquid nitrogen fertilizers. Nitric acid and ammonia are used to make ammonium nitrate (34-0-0), a solid granular fertilizer with a high concentration of nitrogen. The two materials are mixed together in a tank and a neutralization reaction occurs, producing ammonium nitrate. This material can then be stored until it is ready to be granulated and blended with the other fertilizer components.

*Urea (46-0-0)* is a solid nitrogen product typically applied in granular form. It can becombined with ammonium nitrate and dissolved in water to make a highly soluble liquid nitrogen fertilizer known as urea ammonium nitrate (UAN) solution typically containing 28 to 32 % nitrogen.

*Ammonium sulfate (21-0-0)* is another solid nitrogen fertilizer.

### 4.4.4.2. Phosphate fertilizer

Phosphate rock, $Ca_5(PO_4)_3F$ which contains 27 to 38% phosphorus pentoxide ($P_2O_5$), is the main raw material source from which most types of phosphate fertilizers are produced.

In its unprocessed state, phosphate rock is not suitable for direct agricultural appli-cation, since the phosphorus it contains is insoluble. To transform the phosphorus into a plant-available form ($CaH_4(PO_4).H_2O$) and to obtain a more concentrated product, phosphate rock is processed using sulphuric acid, phosphoric acid and/or nitric acid.

Acidulation by means of sulphuric acid converts the rock to monocalcium phosphate popularly known as normal or single superphosphate (SSP) having a phosphorus content of 15-20% $P_2O_5$.

Ground phosphate rock is thoroughly mixed with metered quantities of 60-70% sul-phuric acid in the ratio of 0.82 to 0.95 acid to phosphate rock. The heat of dilution serves to heat the acid to proper reaction temperature. Excess heat is dissipated by evaporation of extra water added. The rate of water and acid addition is varied to control moisture level. The fresh superphosphate drops to a slow moving conveyor in a den where it takes I hour in order to solidify. A disintegrator slices the solid mass of crude product before it is taken to pile storage. The chemical reaction continues for 4-6 weeks to 15-20% $P_2O_5$. After curing, the product is bagged and shipped.

$$2\ Ca_5(PO_4)_3F + 7H_2SO_4 + 3H_2O \longrightarrow 3CaH_4(PO_4)_2.H_2O + 7CaSO_4 + 2HF$$

Sulphuric acid is also used to manufacture phosphoric acid, an intermediate product in the production of triple superphosphate (TSP).

Acidulation of the phosphate rock using phosphoric acid produces triple superphos-phate by the following reaction:

$$Ca_5(PO_4)_3F + 7H_3PO_4 + 5H_2O \longrightarrow 5CaH_4(PO_4)_2.H_2O + HF$$

Since no $CaSO_4$ is formed, the phosphorus content is not diluted. Therefore, TSP (0-46-0) has a phosphorus content of 43-48 percent as $P_2O_5$.

Two processes are used to produce TSP fertilizers: run-of-pile and granular. The run-of-pile process is similar to the SSP process. Granular TSP uses lower strength phosphoric acid (40 percent compared to 50 percent for run-of-pile). Pulverized ground phosphate rock is mixed with phosphoric acid in the reactor. The resultant slurry is sprayed into the granulator. The product is dried, screened and cooled. It is stored for 4-6 weeks to cure.

*Monoammonium Phosphate (MAP) (11-52-0)* and*Diammonium Phosphate (DAP) (18-46-0)* are called ammoniated phosphates because phosphoric acid is treated withammonia to form these basic phosphate products that also contain nitrogen. They are widely produced in the granular form for blending with other types of fertilizers, and are also produced in non-granular forms for use in liquid fertilizers.

Acidulation of phosphate rock using nitric acid produces NP slurries for use in the manufacture of compound fertilizers.

### 4.4.4.3. Potassium fertilizer

Most potassium (K) is obtained from naturally occurring ore deposits. Although the low-grade unrefined mineral ores can be directly applied, it is normally purified.

Potassium in the form of potassium chloride and potassium magnesium sulphate are used in the manufacture of multi-nutrient fertilizers. Potassium chloride is typically supplied to fertilizer manufacturers in bulk. The manufacturer converts it into a more usable form by granulating it. This makes it easier to mix with other fertilizer components.

### 4.4.4.4. Manufacture of NPK fertilizer

The raw materials, in solid form, can be supplied to fertilizer manufacturers in bulk quantities of thousands of tons, drum quantities, or in metal drums and bag containers.

Secondary nutrients are added to some fertilizers to help make them more effective. The different types of particles are blended together in appropriate proportions to produce a composite fertilizer. Typically, complex NPK fertilizers are manufactured by producing slurries of ammonium phosphates, to which potassium salts are added prior to being made into granules. PK fertilizers, on the other hand, are generally produced by granulation of superphosphates (SSP or TSP) with potassium salts.

One method of granulation involves putting the solid materials into a rotating drum which has an inclined axis. (Refer to drum agglomerator in Unit 2). As the drum rotates, pieces of the solid fertilizer take on small spherical shapes. They are passed through a screen that separates out adequately sized particles. A coating of inert dust is then applied to the particles to make them remain discreet and to inhibit moisture retention. Finally, the particles are dried, thus completing the granulation process. The fertilizer is emptied onto a conveyor belt, which transports it to the bagging machine.

### 4.4.5. Quality Control

To ensure the quality of the fertilizer that is produced, manufacturers monitor the product at each stage of production. The raw materials and the finished products are all subjected to a battery of physical and chemical tests to show that they meet the specifications previously developed. Some of the characteristics that are tested include pH, appearance, density, and melting point. Since fertilizer production is government-regulated, tests are run on samples to determine total content of nitrogen, phosphate, and other elements in the specifications.

## 4.5. Manufacture Of Portland Cement

### 4.5.1. Introduction

Historically, cement can be traced back to the early Roman Empire. It contributed to the building of the great structures of the Roman Empire. Portland cement is a fine powder, generally gray in colour. It is composed primarily of calcium silicates, calcium aluminates, and calcium ferrites. When mixed with water (hydrated), cement solidifies to an artificial rock, similar to Portland stone. A Portland stone is a yellow limestone from the Isle of Portland, in Great Britain. Hence the name Portland cement. By varying the amounts and types of the same basic ingredients, cement with various properties may be obtained. Concrete is a mixture of gravel, sand and cement.

### 4.5.2. Raw materials

The major components of cement in terms of metal oxides are $CaO$, $SiO_2$, $Al_2O_3$, and $Fe_2O_3$. Typically, Ca is provided from limestone, Si from sand or flyash, Al from flyash or clay, and Fe from iron ore or slag.

### 4.5.3. Manufacturing process

Fig 4.8 is a process flow diagram for a typical cement manufacturing plant.

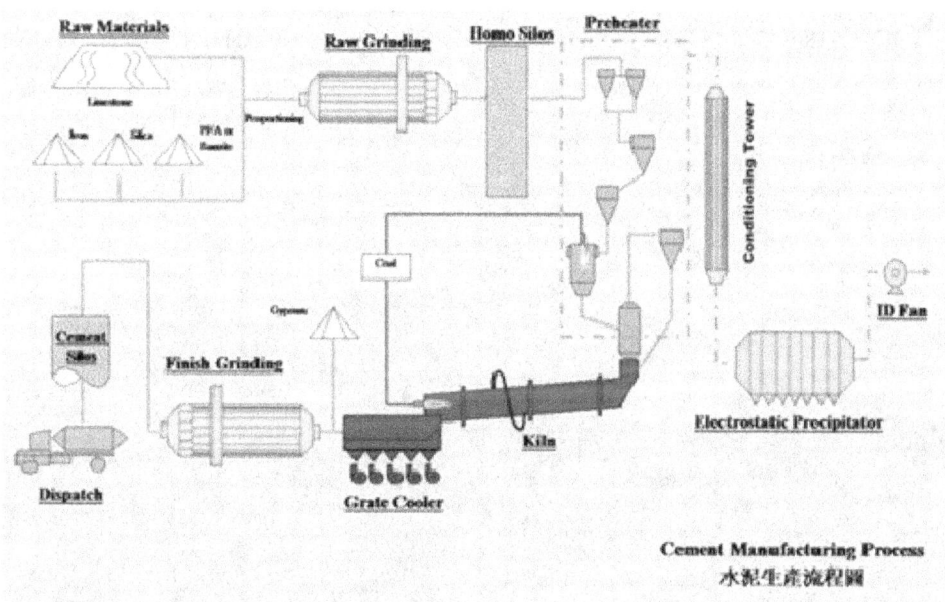

**Fig 4.8.** A cement manufacturing process

## 5.1.7. Grinding

The feed to the grinding process is proportioned to meet a desired chemical composition. Typically, it consists of 80% limestone, 9% silica, 9% flyash, and 2% iron ore.

These materials are ground to 75 micron in a ball mill. Grinding can be either wet or dry. The "raw meal" from dry milling is stored in a homogenizing silo in which the chemical variation is reduced. In the wet process, each raw material is fed with water to the ball mill. This slurry is pumped to blending tanks and homogenized to correct chemical composition. The slurry is stored in tanks until required.

## 4.5.3.2. Pyroprocessing

In the preheater, the raw meal from the mill is heated with the hot exhaust gas from the kiln before being fed into the rotary kiln to form a semi-product known as clinker. The ash from fuel used is also absorbed into the clinker. The particle size range for clinker is from about 2 inches to about 10 mesh.

ILLUSTRATION OF SIMPLE CEMENT KILN

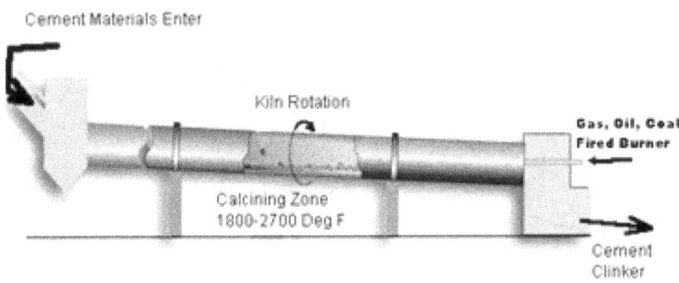

**Fig 4.9.** A cement kiln

## 4.5.3.3. Reactions in the kiln

Basic chemical reactions are: evaporating all moisture, calcining the limestone to produce free calcium oxide, and reacting the calcium oxide with the minor materials(sand, shale, clay, and iron). This results in a final black, nodular product known as "clinker" which has the desired hydraulic properties.

A summary of the physical and chemical reactions that take place in the kiln are shown in Table 4.1.

## Table 4.1. Reactions in a cement kiln

| T $^{o}C$ | Reaction | Remarks |
|---|---|---|
| 100 | Evaporation of water | Solid phase reactions, endo-Thermic |
| >500 | Evolution of combined water from the clay | Solid phase reactions, endo-Thermic |
| 900 | Crystallization of amorphous dehydration prodUcts<br><br>Carbon dioxide evolution from $CaCO_3$ | Solid phase reactions, endo-Thermic |
| 900 -1200 | Main reactions between lime and clay to form | Fusion reactions, exothermic |

| | | | |
|---|---|---|---|
| | Clinker | | |
| 1250 - 1280 | Beginning of liquid for-Mation | | Fusion reactions, endothermic |
| 1280 -1550 | Further liquid formation and final cement forma-Tion | | Fusion reaction, endothermic |

The main reactions which give the real strength of cement are as follows:

$$2CaO + CaO.SiO_2 \xrightarrow{900-1200} CaO + 2CaO.SiO_2 \xrightarrow{1200-1500} 3CaO.SiO_2$$

The main constituents of clinker are shown in Table 4.2.

## Table 4.2. Main constituents of clinker

| | Abbreviation | Common name | Function |
|---|---|---|---|
| $2CaO.SiO_2$ | $C_2S$ | Dicalcium silicate | Together with $3CaO.SiO_2$, responsible for final strength (I year) |
| $3CaO.SiO_2$ | $C_3S$ | Tricalciumsili-cate | Responsible for early strength i.e. 7-8 days |
| $3CaO.Al_2O_3$ | $C_3A$ | Tricalciumalumi-nate | Causes fast harden-ing; needs retarda-tion by gypsum by forming $3CaO.Al_2O_3CaSo_4.3H_2O$ |
| $3CaO. Al_2O_3. Fe_2O_3$ | $C_4AF$ | Tetracalcium alumino-ferrate | Improves chemical resistance |

The rotary kiln discharges the red-hot clinker under the intense flame into a clinker cooler. The clinker cooler recovers heat from the clinker and returns the heat to the pyroprocessing system thus reducing fuel demand and cooling the clinker to a temperature conducive for handling in subsequent steps.

### 4.5.3.4. Finish Grinding

The final process of cement making is called finish grinding. The clinker is dosed with a controlled amount of gypsum and fed into a finish mill. Other additives may be added during the finish grinding process to produce formulated cements such as waterproofing and corrosion resistant cements.

The cement is stored in a bulk silo for packaging and/or bulk distribution.

## Unit objectives

At the end of this unit you should be able to:

a. Describe using equations and diagrams, the electrolytic production of sodium hydroxide and chlorine using mercury, diaphram and membrane cells

b. Explain how ammonia is manufactured from methane and air by the Haber process

c. Describe the Contact process for the manufacture of sulphuric acid

d. Discuss the various types of fertlizers and the manufacture of phosphate fertilizer

e. Describe using diagrams, equations and unit operations, the process for the manufacture of Portland cement.

## Summary of the Chapter

In this Chapter, we will look at industrial manufacture of some of the common basic chemicals. You will learn about the processes and the chemistry involved in the manufacture of, sodium hydroxide and chlorine, ammonia, sulphuric acid, fertilizer and portland cement.

## References

1. Chang R. and Tikkanen W. (1988). The Top fifty Industrial

Chemicals.

2. George T. A. (1977). *Shreve's Chemical Process Industries.* *5ᵗʰedn.*McGRAW-HILL  INTERNATIONAL  EDITIONS. Chemical Engineering Series. Singa-pore.

3. Shukla S. D and Pandey G. N, (1978). A Textbook of Chemical Technology. Vol.1 (Inorganic/Organic). Vikas publishing House PVT Ltd. New Delhi

4. Stephenson R.M. (1966). Introduction to the Chemical Process Industries, Reinhold Publishing Corporation, New York.

5. Gerhartz, W. (Editor), (1987). Ullmann's Encyclopaedia of Industrial Che-mistry, 5ᵗʰ Edition, VCH VerlagsgesellschaftmbH, Weinheim.

**List of relevant resources**

- Computer with internet facility to access links and relevant copywrite free resources
- CD-Rom accompanying this module for compulsory reading and demonstra-tions
- Multimedia resources like video,VCD and CDplayers

**List of relevant useful links**

http://cheresources.com

http:/uk.encarta.msn.com/media_761566936/Sulphuric

_Acid.html http/www.icis.com/chemical/intelligence

These sites have information and other links on manufacturing technologies of various chemical products.

# CHAPTER – 5

---

# Organic Chemical Industries: Petroleum, Petrochemicals andPolymers

☞CONTENTS

**5.1 Petroleum processing**

**5.2 Petrochemicals**

**5.3 Polymers**

## 5.1 Petroleum Processing

## 5.1.1. Introduction

The term *petroleum* comes from the Latin stems *petra*, "rock," and *oleum*, "oil." It is used to describe a broad range of hydrocarbons that are found as gases, liquids, or solids beneath the surface of the earth. The two most common forms are natural gas and crude oil.

*Natural gas:* Natural gas which is a mixture of lightweight alkanes, accumulates inporous rocks. A typical sample of natural gas when it is collected at its source contains about 80% methane ($CH_4$), 7% ethane ($C_2H_6$), 6% propane ($C_3H_8$), 4% butane and isobutane ($C_4H_{10}$), and 3% pentanes ($C_5H_{12}$). The $C_3$, $C_4$, and $C_5$ hydrocarbons are removed before the gas is sold. The commercial natural gas delivered to the customer is therefore primarily a mixture of methane and ethane. The propane and butanes removed from natural gas are usually liquefied under pressure and sold as liquefied petroleum gases (*LPG*).

***Crude oil*** is a composite mixture of hydrocarbons (50-95% by weight) occurringnaturally. The first step in refining crude oil involves separating the oil into different hydrocarbon fractions by distillation. Each fraction is a complex mixture. For example, more than 500 different hydrocarbons can be found in the gasoline fraction.

Petroleum is found in many parts of the world which include the Middle East, southern United States, Mexico, Nigeria and the former Soviet Union.

### 5.1.2. Uses of petroleum

Most of the crude oil is used in the production of fuels such as gasoline, kerosene, and fuel oil. Non-fuel uses include petroleum solvents, industrial greases and waxes, or as raw materials for the synthesis of *petrochemicals*. Petroleum products are used to produce synthetic fibres such as nylon and other polymers such as polystyrene, polyethylene and synthetic rubber. They also serve as raw materials in the production of refrigerants, aerosols, antifreeze, detergents, dyes, adhesives, alcohols, explosives and pesticides. The $H_2$ given off in refinery operations can be used to produce a number of inorganic petrochemicals, such as ammonia, ammonium nitrate, and nitric acid from which most fertilizers as well as other agricultural chemicals are made.

### 5.1.3. Oil Extraction

The vast majority of petroleum is found in oilfields or reservoirs below the earth's surface as shown in Fig 5.1.

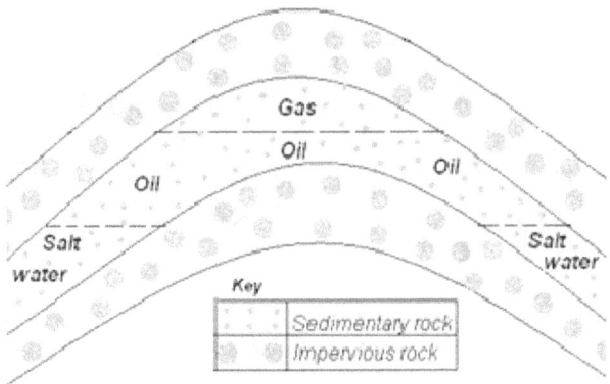

**Fig 5.1**   Schematic diagram of a crude oil reservoir

The oil is sometimes under high pressure and can flow to the surface on its own without pumping. However, most wells require induced pressure using water, carbon dioxide, natural gas or steam in order to bring the oil to the surface.

Petroleum refining has evolved continuously in response to changing consumer demand for better and different products. The original requirement was to produce kerosene as a cheaper and better source of light than whale oil. The development of the internal combustion engine led to the production of gasoline and diesel fuels. The evolution of the airplane created an initial need for high-octane aviation gasoline and then for jet fuel, a sophisticated form of the original product, kerosene. Present-day refineries produce a variety of products including many required as feedstock for the petrochemical industry. Common petroleum products include gasoline, liquefied refinery gas, still gases, kerosene, aviation fuel, distillate fuel oil, residual fuel oil, lubricating oils, asphalt, coke and petrochemical feedstocks.

The history of petroleum refining is given in Table 5.1.

## Table 5.1. History of Petroleum refining

| Year | Process | Purpose | By-Products, etc. |
|------|---------|---------|-------------------|
| 1862 | Atmospheric distil-Lation | Produce kerosene | Naphtha, tar, etc. |
| 1870 | Vacuum distillation | Lubricants originally, then crackingfeedstocks (1930's) | Asphalt, residual, Coker feedstocks |
| 1913 | Thermal cracking | Increase gasoline yield | Residual, bunker fuel |
| 1916 | Sweetening | Reduce sulfur&odor | Sulfur |
| 1930 | Thermal reforming | Improve octane number | Residual |
| 1932 | Hydrogenation | Remove sulfur | Sulfur |
| 1932 | Coking | Produce gasoline basestock | Coke |
| 1933 | Solvent extraction | Improve lubricant viscosity index | Aromatics |
| 1935 | Solvent dewaxing | Improve pour point | Waxes |
| 1935 | Cat. polymeriza-tion | Improve gasoline yield and octane number | Petrochemical, feedstocks |
| 1937 | Catalytic cracking | Higher octane gasoline | Petrochemical, feedstocks |
| 1939 | Visbreaking | Reduce viscosity | Increased distillate, tar |
| 1940 | Alkylation | Increase gasoline octane & yield | High-octane aviation gasoline |
| 1940 | Isomerization | Produce alkylation feedstock | Naphtha |
| 1942 | Fluid catalytic cracking | Increase gasoline yield & octane Petrochemical feedstocks | |
| 1950 | Deasphalting | Increase cracking feedstock | Asphalt |
| 1952 | Catalytic reform-ing | Convert low-quality naphtha | Aromatics |
| 1954 | Hydrodesulfuriza-tion | Remove sulfur | Sulfur |

| 1956 | Inhibitor sweetening | Remove mercaptan | Disulfides |
|------|----------------------|------------------|------------|
| 1957 | Catalytic isomerization | Convert to molecules with high octane number | Alkylation feed-stocks |
| 1960 | Hydrocracking | Improve quality and reduce sulfur | Alkylation feed-stocks |
| 1974 | Catalytic dewaxing | Improve pour point | Wax |
| 1975 | Residual hydro-cracking | Increase gasoline yield from residual | Heavy residuals |

Source: http://www.setlaboratories.com

## 5.1.4. Characteristics and classification of Crude Oil

As has been mentioned, crude oils are complex mixtures containing many different hydrocarbon compounds that vary in appearance and composition from one oil field to another. Crude oils range in consistency from water to tar-like solids, and in colour from clear to black.

An "average" crude oil contains about 84% carbon, 14% hydrogen, 1%-3% sulfur, and less than 1% each of nitrogen, oxygen, metals, and salts. Crude oils are generally classified as paraffinic, naphthenic, or aromatic, based on the predominant proportion of similar hydrocarbon molecules. Oils with low carbon, high hydrogen, and high

API (American Petroleum Institute) gravity are usually rich in paraffins and tend to yield greater proportions of gasoline and light petroleum products; those with high carbon, low hydrogen, and low API gravities are usually rich in aromatics. The former category is known as light crudes and the latter as heavy crudes.

Crude oils that contain appreciable quantities of hydrogen sulfide or other reactive sulfur compounds are generally called "sour." while those with lower sulfur are called "sweet."

### 5.1.4. Composition of petroleum

**Crude petroleum contain hydrocarbon and non-hydrocarbon compounds. Hyrocarbon compounds**

*Paraffins*-The paraffinic crude oil hydrocarbon compounds found in crude oil havethe general formula $C_nH_{2n+2}$ and can be either straight chains (normal) or branched chains (isomers) of carbon atoms. The lighter, straight chain paraffin molecules are found in gases and paraffin waxes. The branched-chain (isomer) paraffins such as isobutene are usually found in heavier fractions of crude oil and have higher octane numbers than normal paraffins.

*Aromatics:*The aromatic series include**simple aromatic compounds such as benzene, naphthalenes and the most complex aromatics, the polynuclears**whichhave three or more fused aromatic rings. They have high anti-knock value and good storage stability.

*Naphthenes (Naphtha):* **These**are saturated hydrocarbon groupings with the generalformula $C_nH_{2n}$, arranged in the form of closed rings (cyclic) and found in all fractions of crude oil except the very lightest. Single-ring naphthenes (monocycloparaffins) with five and six carbon atoms such as cyclohexane predominate. Two-ring naphthenes (dicycloparaffins) are found in the heavier ends of naphtha.

*Alkenes (Olefins):* Olefins such as ethylene, butene, isobutene are usually formed bythermal and catalytic cracking and rarely occur naturally in unprocessed crude oil. They are unstable and also improve the anti-knock tendencies of gasoline but not as much as the iso-alkanes. When stored, the olefins polymerise and oxidize. This tendency to react is employed in the production of petrochemicals.

*Dienes and Alkynes:*Examples of dienes or diolefins, are 1,2-butadiene and 1,3-butadiene. Acetylene is a typical alkyne. This category of hydrocarbons is obtained from lighter fractions through cracking.

## Non-hydrocarbons

*Sulfur Compounds:*Sulfur may be present in crude oil as hydrogen sulfide ($H_2S$), asmercaptans, sulfides, disulfides, thiophenes, etc. or as elemental sulfur. Each crude oil has different amounts and types of sulfur compounds, but as a rule the proportion, stability, and complexity of the compounds are greater in heavier crude-oil fractions. Sulphur is an undesirable component because of its strong offensive odour, corrosion, air pollution by some of its compounds and its effect of reducing tetraethyl lead

(anti-knock agent). Hydrogen sulfide is a primary contributor to corrosion in refinery processing units. Other corrosive substances are elemental sulfur and mercaptans. The corrosive sulfur compounds also have an obnoxious odor. The combustion of petroleum products containing sulfur compounds produces undesirables such as sulfuric acid and sulfur dioxide. Catalytic hydrotreating processes such as hydrode-sulfurization remove sulfur compounds from refinery product streams. Sweetening processes either remove the obnoxious sulfur compounds or convert them to odorlessdisulfides, as in the case of mercaptans.

***Oxygen Compounds:***Oxygen compounds such as phenols, ketones, and carboxylicacids occur in crude oils in varying amounts.

***Nitrogen Compounds:***Nitrogen is found in lighter fractions of crude oil as basiccompounds, and more often in heavier fractions of crude oil as nonbasic compounds. Nitrogen oxides can form in process furnaces. The decomposition of nitrogen compounds in catalytic cracking and hydrocracking processes forms ammonia and cyanides that can cause corrosion.

***Trace Metals:***Metals, including nickel, iron, and vanadium are often found in crudeoils in small quantities and are removed during the refining process. Burning heavy fuel oils in refinery furnaces and boilers can leave deposits of vanadium oxide and nickel oxide in furnace boxes, ducts, and tubes. It is also desirable to remove trace amounts of arsenic, vanadium, and nickel prior to processing as they can poison certain catalysts.

***Salts:*** Crude oils often contain inorganic salts such as sodium chloride, magnesiumchloride, and calcium chloride in suspension or dissolved in entrained water (brine) in the form of an emulsion. These salts must be removed or neutralized before processing to prevent catalyst poisoning, equipment corrosion, and fouling. Salt corrosion is caused by the hydrolysis of some metal chlorides to hydrogen chloride (HCl) and the subsequent formation of hydrochloric acid when crude oil is heated. Hydrogen chloride may also combine with ammonia to form ammonium chloride ($NH_4Cl$), which causes fouling and corrosion. Salt is removed mainly by mechanical or electrical desalting

***Carbon Dioxide***:Carbon dioxide may result from the decomposition of bicarbonatespresent in or added to crude, or from steam used in the distillation process.

***Naphthenic Acids***:Some crude oils contain naphthenic (organic) acids, which maybecome corrosive at temperatures above 230°C when the acid value of the crude is above certain level.

## 5.1.5.  Petroleum Refining

The petroleum industry began with the successful drilling of the first commercial oil well in 1859, and the opening of the first refinery two years later to process the crude into kerosene. Today, petroleum refinery products obtained include gasoline, kerosene, propane, fuel oil, lubricating oil, wax, and asphalt.

Refining crude oil involves two kinds of processes: First, there are physical proces-ses which simply refine the crude oil (without altering its molecular structure) into useful products such as lubricating oil or fuel oil. Petroleum refining begins with **distillation**, or fractionation, which separates crude oil in atmospheric and vacuumdistillation towers into groups of hydrocarbon compounds of differing boiling-point ranges called "fractions" or "cuts."

Second, there are **chemical conversion processes** which alter the size and/or mole-cular structure of hydrocarbon molecules to produce a wide range of products, some of them known by the general term petrochemicals. Conversion processes include:

- **Decomposition** (dividing) by thermal and catalytic cracking;

- **Unification** (combining) through alkylation and polymerization; and

- **Alteration** (rearranging) with isomerization and catalytic reforming.

As seen above, the major chemical conversions include cracking, alkylation, polymerisation, isomerisation and reforming. The converted products are then subjected to various treatment and separation processes.

**Treatment Processes** are intended to prepare hydrocarbon streams for additionalprocessing and to prepare finished products. Treatment may include the removal or separation of aromatics and naphthenes as well as impurities and undesirable contaminants. Treatment may involve chemical or physical separation such as dissolving, absorption, or precipitation using a variety and combination of processes including hydrodesulfurizing and sweetening.

**Formulating and Blending** is the process of mixing and combining hydrocarbonfractions, additives, and other components to produce finished products with specific performance properties. Integrated refineries incorporate fractionation, conversion, treatment, and blending operations and may also include petrochemical processing.

## 5.1.6. Octane number and the development of cracking and reforming processes

About 10% of the product of the distillation of crude oil is a fraction known as **straight-run gasoline**, which served as a satisfactory fuel during the early days ofthe internal combustion engine. As the automobile engine developed, it was made more powerful by increasing the compression ratio. Modern cars run at compression ratios of about 9:1, which means the gasoline-air mixture in the cylinder is compressed by a factor of nine before it is ignited. Straight-run gasoline burns unevenly in high-compression engines, producing a shock wave that causes the engine to "knock," The challenge for the petroleum industry was to increase the yield of gasoline from each barrel of crude oil and to decrease the tendency of gasoline to knock when it burned. It was found that:

- Branched alkanes and cycloalkanes burn more evenly than straight-chain alkanes.
- Short alkanes ($C_4H_{10}$) burn more evenly than long alkanes ($C_7H_{16}$).
- Alkenes burn more evenly than alkanes.
- Aromatic hydrocarbons burn more evenly than cycloalkanes.

The most commonly used measure of a gasoline's ability to burn without knocking is its **octane number**. Octane numbers compare a gasoline's tendency to knock against the tendency to knock of a blend of two hydrocarbons heptane and 2,2,4-trimethyl-pentane, (isooctane). Heptane produces a great deal of knocking while isooctane is more resistant to knocking. Gasolines that match a blend of 87% isooctane and 13% heptane are given an octane number of 87.

There are three ways of reporting octane numbers. Measurements made at high speed and high temperature are reported as *motor octane numbers* while measurements taken under relatively mild engine conditions are known as *research octane numbers*. The *road-index octane numbers* reported on gasoline pumps are an average of these two. Road-index octane numbers for a few pure hydrocarbons are given in the Table 5.2.below.

**Table 5.2. Hydrocarbon Road Octane Numbers**

| Hydrocarbon | Road Index Octane Number |
|---|---|
| Heptane | 0 |
| 2-Methylheptane | 23 |
| Hexane | 25 |
| 2-Methylhexane | 44 |
| 1-Heptene | 60 |
| Pentane | 62 |
| 1-Pentene | 84 |
| Butane | 91 |
| Cyclohexane | 97 |
| 2,2,4-Trimethylpentane (iso-octane) | 100 |
| Benzene | 101 |
| Toluene | 112 |

By 1922 a number of compounds had been discovered that could increase the octane number of gasoline. Adding as little as 6 ml of tetraethyllead to a gallon of gasoline, for example, can increase the octane number by 15 to 20 units. This discovery gave rise to the first «ethyl» gasoline, and enabled the petroleum industry to produce aviation gasolines with octane numbers greater than 100.

Another way to increase the octane number is **thermal reforming**. At high tempera-tures (500-600°C) and high pressures (25-50 atm), straight-chain alkanes isomerize to form branched alkanes and cycloalkanes, thereby increasing the octane number of the gasoline. Running this reaction in the presence of hydrogen and a catalyst such as a mixture of silica ($SiO_2$) and alumina ($Al_2O_3$) results in **catalytic reforming**, which can produce a gasoline with even higher octane numbers.

The yield of gasoline is increased by "cracking" the long chain hydrocarbons into smaller pieces at high temperatures (500°C) and high pressures (25 atm). A saturated $C_{12}$ hydrocarbon in kerosene, for example, might break into two $C_6$ fragments. Because the total number of carbon and hydrogen atoms remains constant, one of the products of this reaction must contain a C=C double bond.

$$CH_3(CH_2)_{10}CH_3 \longrightarrow CH_3(CH_2)_4CH_3 + CH_2=CH(CH_2)_3CH_3$$

The presence of alkenes in thermally cracked gasolines increases the octane number (70) relative to that of straight-run gasoline (60), but it also makes thermally-crac-ked gasoline less stable for long-term storage. Thermal cracking has therefore been replaced by **catalytic cracking**, which uses catalysts instead of high temperatures and pressures to crack long-chain hydrocarbons into smaller fragments for use in gasoline.

## 5.1.8.  Catalytic Cracking

Ethylene and propylene are the most important organic chemical feedstocks accounting for over 50-60% of all organic chemicals. But because of their relatively high reactivities, very few olefins are found in natural gas or crude oil. Therefore, they must be manufactured by cracking processes.

The purpose of cracking is to break complex hydrocarbons into simpler molecules in order to increase the quality and quantity of lighter, more desirable products and decrease the amount of residuals. The heavy hydrocarbon feedstock is cracked into lighter fractions such as kerosene, gasoline, LPG, heating oil, and petrochemical feedstock. LPG gases are feedstock for olefins such as ethylene and propylene.

The decomposition takes place by catalytic action or heating in the absence of oxygen

(pyrolysis). The catalysts used in refinery cracking units are typically zeolite, aluminumhydrosilicate, treated bentonite clay, fuller's earth, bauxite, and silica-alumina ($SiO_2$-$Al_2O_3$) all of which come in the form of powders, beads, or pellets.

The formation of gasoline (with low molecular weight) from heavy gas oil of high molecular weight is shown in the following equation:

$$(C_7H_{15})_2(CH_2)_{15} \xrightarrow{\text{Heat}} C_7H_{16} + C_6H_{12}:CH_2 + C_{14}H_{28}:CH_2$$

| Heavy gas oil | | Gasoline | Gasoline (antiknock) | Recycle stock |

There are three basic functions in the catalytic racking process:

- Reaction - Feedstock reacts with catalyst and cracks into different hydrocarbons

- Regeneration - Catalyst is reactivated by burning off coke

- Fractionation - Cracked hydrocarbon stream is separated into various products.

## 5.1.8.   Catalytic Reforming

Catalytic reforming is an important process used to convert low-octane naphthas into high-octane gasoline blending components called reformates. Depending on the properties of the naphtha feedstock (as measured by the paraffin, olefin, naphthene, and aromatic content) and catalysts used, reformates can be produced with very high concentrations of toluene, benzene, xylene, and other aromatics useful in gasoline blending and petrochemical processing. Hydrogen, a significant by-product, is separated from the reformate for recycling and use in other processes. Most processes use platinum as the active catalyst. Sometimes platinum is combined with a second catalyst (bimetallic catalyst) such as rhenium or another noble metal The conversion is illustrated by the following reaction in which a cycloalkane is converted to an aromatic compound, usually of higher octane number.

The naphtha feed is mixed with recycled hydrogen and introduced to the feed preheater to raise the temperature. The hot mixture of hydrogen and naphtha vapours is passed through a series of four reactors containing the catalyst. The working temperatures and pressures are usually between 150oC to 510oC and 1500kPa to 7000kPa respectively. The products are cooled and about 90% of the hydrogen is compressed and recycled. The main product is fractionated. The overhead product can be used as a fuel.

Some catalytic reforming systems continuously regenerate the catalyst while in other systems one reactor at a time is taken off-stream for catalyst regeneration, Some facilities regenerate catalysts from all of the reactors during shutdown.

## 5.1.9. Polymerization

This is the joining up of low molecular weight fractions to form high molecular weight components. This process converts by-products hydrocarbon gases into liquid hydrocarbons that are suitable for use as high octane number fuels and for petrochemical industry. The combining molecules are usually unsaturated. Propylene and iso-butylene are common olefins polymerised in the vapour phase in reactions such as shown here:

## 5.1.10. Alkylation

Akylation was discussed in Unit 2 under unit processes. In the context of petroleum processing, it is the combining of an olefin with an aromatic hydrocarbon. The process is of relatively minor importance compared to the catalytic cracking and catalytic re-forming processes.

It is used mainly for converting gaseous hydrocarbons to gasoline in the presence of an acid catalyst such as hydrogen fluoride or sulphuric acid. The processes are usually exothermic and similar to polymerisation. An example is the formation of 2,2-Dimethylbutane from ethylene and isobutane:

$$C{=}C \;+\; C{-}\overset{\displaystyle |}{\underset{\displaystyle C}{C}}{-}C \;\longrightarrow\; C{-}\overset{\displaystyle C}{\underset{\displaystyle C}{\overset{|}{\underset{|}{C}}}}{-}C{-}C$$

Ethylene

## 5.1.11. Treating/Sweetening drying

Treating is a means by which contaminants such as organic compounds containing sulfur, nitrogen, and oxygen; dissolved metals and inorganic salts; and soluble salts dissolved in emulsified water are removed from petroleum fractions or streams. Swee-tening, a major refinery treatment of gasoline, treats sulfur compounds (hydrogen sulfide, thiophene and mercaptan) to improve colour, odour, and oxidation stability.

Sweetening also reduces concentrations of carbon dioxide.

Treating can be accomplished at an intermediate stage in the refining process, or just before sending the finished product to storage. Choices of a treating method depend on the nature of the petroleum fractions, amount and type of impurities in the fractions to be treated, the extent to which the process removes the impurities, and end-product specifications. Treating materials include acids, solvents, alkalis, oxidizing, and adsorption agents.

Sulfuric acid is the most commonly used acid treating process. Sulfuric acid treating results in partial or complete removal of unsaturated hydrocarbons, sulfur, nitrogen, oxygen compounds, resinous and asphaltic compounds. It is used to improve the odour, colour, stability, carbon residue, and other properties of the oil. Clay/lime treatment of acid-refined oil removes traces of asphaltic materials and other com-pounds to improve product colour, odour, and stability. Caustic treating with sodium (or potassium) hydroxide is used to improve odour and colour by removing organic acids (naphthenic acids, phenols) and sulfur compounds (mercaptans, $H_2S$) by a caustic wash. By combining caustic soda solution with various solubility promoters (e.g., methyl alcohol and cresols), up to 99% of all mercaptans as well as oxygen and nitrogen compounds can be dissolved from petroleum fractions.

*Drying* is accomplished by the use of water absorption or adsorption agents to re-move water from the products. Some processes simultaneously dry and sweeten by adsorption on molecular sieves.

## Sulfur Recovery

Sulfur recovery converts hydrogen sulfide in sour gases and hydrocarbon streams to elemental sulfur. A typical process produces elemental sulfur by burning hydrogen sulfide under controlled conditions. The gases are then exposed to a catalyst to recover additional sulfur. Sulfurvapor from burning and conversion is condensed and recovered.

## Hydrogen Sulfide Scrubbing

Hydrogen sulfide scrubbing is a common treating process in which the hydrocarbon feedstock is first scrubbed to prevent catalyst poisoning. Hydrotreating for sulfur removal is called **hydrodesulfurization**. In a typical catalytic hydrodesulfurization unit, the feedstock is deaerated and mixed with hydrogen, preheated in a fired heater (315°-425°C) and then charged under pressure (up to 1,000 psi) through a fixed-bed catalytic reactor. In the reactor, the sulfur and nitrogen compounds in the feedstock are converted into $H_2S$ and $NH_3$. The reaction products leave the reactor and after cooling to a low temperature enter a liquid/gas separator. The hydrogen-rich gas from the high-pressure separation is recycled to combine with the feedstock, and the low-pressure gas stream rich in $H_2S$ is sent to a gas treating unit where $H_2S$ is removed.

The clean gas is then suitable as fuel for the refinery furnaces. The liquid stream is the product from hydrotreating and is normally sent to a stripping column for removal of $H_2S$ and other undesirable components. In cases where steam is used for stripping, the product is sent to a vacuum drier for removal of water. Hydrodesulfurized products are blended or used as catalytic reforming feedstock.

## 5.2. Petrochemical Industry

Petrochemicals are chemicals, other than fuels, derived from petroleum. These chemicals include a large number of aliphatic and aromatic organic compounds of various functional groups. Examples include benzene and it derivatives, methane, ethylene, propylene, butene, toluene, and xylene and their derivatives.

We shall study the manufacture of two important petrochemicals, namely pthalic anhydride and adipic acid.

## 5.2.1. Phthalic Anhydride

## 5.2.1.1 Introduction

Phthalic anhydride ($C_6H_4(CO)_2O$) is a colourless solid which was first reported in 1836. It is a precursor to a variety of industrial organic chemicals.

## 5.2.1.2. Uses

Phthalic anhydride is a versatile intermediate in organic chemistry partly because it is bifunctional and cheaply available. Most characteristically, it undergoes hydrolysis and alcoholysis. Hydrolysis by hot water, forms *ortho*-phthalic acid. This process is reversible with phthalic anhydride re-forming upon heating the acid above 180 °C.

The alcoholysis reaction is the basis of the manufacture of phthalate esters which are widely used plasticizers. These are additives that give polymers more flexibility.

Reaction of phthalic anhydride with alcohols gives diesters as follows:

$$C_6H_4(CO)_2O + ROH \longrightarrow C_6H_4(CO_2H)CO_2R$$

The second esterification is more difficult and requires removal of water:

$$C_6H_4(CO_2H)CO_2R + ROH \longrightarrow C_6H_4(CO_2R)_2 + H_2O$$

Two of the most important diestersbis(2-ethylhexyl) phthalate (DEH) and dioctyl phthalate (DOP) are used as plasticisers in the manufacture of polyvinyl chloride.

Other major uses of phthalic anhydride are in polyester resins and (decreasingly) in alkyd resins. Alkyd resins containing PA are used in solvent-borne protective coatings. As paint technology increasingly utilises water-borne technologies, many PA-based alkyds have lost out to alternative raw materials. Phthalic anhydride is widely used in industry for the production of certain dyes such as the well-known anthroquinone dye.

### 5.2.1.3. Production Process

Phthalic anhydride was first produced by the oxidation of naphthalene from coal in concentrated sulphuric acid in the presence of mercury sulphate. This route was later replaced by the catalytic vapour phase oxidation of naphthalene in air in the presence of a vanadium oxide catalyst. Today, naphthalene feedstock has been generally superseded by the use of orthoxylene obtained from refineries and crackers. The com-monly used catalyst is vanadium pentoxide with titanium dioxide-antimony trioxide. Alternative catalysts include molybdenum trioxide and calcium oxide, or manganese oxides. The respective reactions for the two feedstocks are as follows:

$$C_{10}H_8 + 4.5O_2 \longrightarrow C_6H_4(CO)_2O + 2H_2O + 2CO_2$$

$$C_6H_4(CH_3)_2 + 3O_2 \longrightarrow C_6H_4(CO)_2O + 3H_2O \quad \Delta H_{298} = -1,127 \text{ kJ/mol}$$

The process technology has changed little although yields have improved and catalysts in current use have a longer life of three years. Another development has been the lowering of the air to orthoxylene weight ratio to 9.5:1, down from about 20:1, thus allowing a reduction in capital costs and energy savings.

In the orthoxylene-based process shown in Fig 5.4, the feedstock is vapourized and mixed with air. The air-orthoxylene mixture is fed to a reactor with vertical tubes filled with catalyst. The reaction takes place at 375-425°C and below 1 bar pressure.

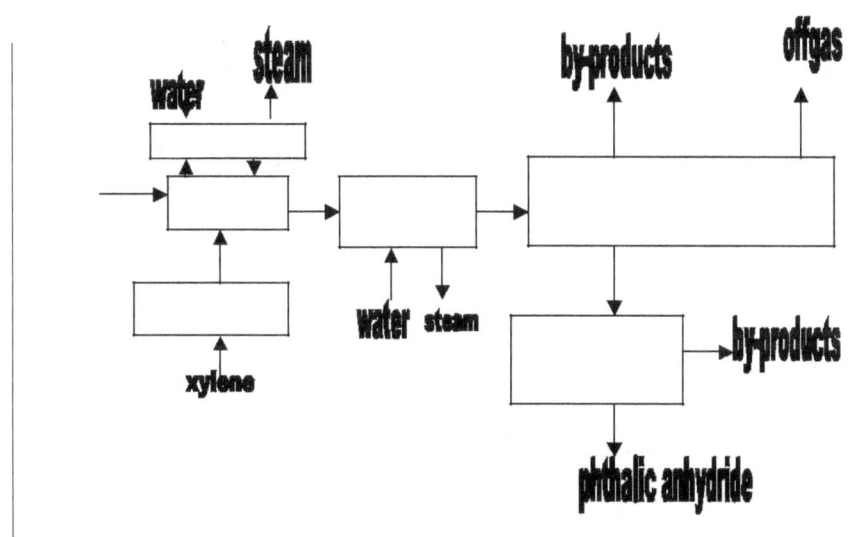

**Fig 5.4**  Process flow diagram for the manufacture of phthalic anhydride.

Temperature control is important because the main reaction and the side reactions are exothermic. The reactor is cooled by recirculating a molten salt on the outside of the tubes.

About 70% of xylene is converted to PA, 15 % is unconverted, 15% is oxidized into maleic anhydride while the 1% balance forms a heavy impurity. The maleic anhydride is formed by the following reaction:

$$C_6H_4(CH_3)_2 + 7.5O_2 \longrightarrow C_4H_2O_3 + 4H_2O + 4CO_2$$

The gases from the reactor are cooled before entering a series of three switch condensers which separates phthalic anhydride from the byproducts. Because of the low partial pressure of phthalic anhydride in the stream, it desublimates rather than condenses.

Phthalicanhydride collects on the walls of the condensers as a solid. When the load on the heat transfer surface reaches a certain level, gas flow is stopped and higher temperature oil is circulated in the tubes to melt the solid. When one condenser is on desublimation mode, the second is in the melting mode while the third is on standby. Purification of the product is by vacuum distillation. A purified product with a purity of over 99.5 % phthalic anhydride is obtained. The product is stored either in a molten state or bagged as flakes.

## 5.2.2. Adipic Acid

### 5.2.2.1. Introduction

Adipic acid, $COOH-(CH_2)_3-COOH$, is one of the most important commercially available aliphatic dicarboxylic acids. The two -COOH groups can yield two kinds of salts. Typically, it is a white crystalline solid,slightly soluble in water and soluble in alcohol and acetone.

Its consumption and production is dominated by the United States. Of the 2.3 million metric tons of adipic acid produced worldwide, about 42% is produced in the United States which also consumes 62% of total production. Western Europe produces about

40%, Asia-Pacific 13%, while the other regions account for the remaining 5%.

### 5.2.2.2. Uses

Adipic acid consumption is linked almost 90% to nylon. Nylon is used for everyday applications such as electrical connecters, cable tires, fishing lines, fabrics, carpeting, and many other useful products. Technical grade adipic acid is used to make plasti-cizers, unsaturated polyesters for production of rigid and flexible resins and foams, wire coatings, elastomers, adhesives and lubricants. Food grade adipic acid is used as a gelling aid, acidulant, leavening and buffering agent.

### 5.2.2.3. Production Process

Almost all of the commercial adipic acid is produced from cyclohexane in a continuous process as shown in Fig. 5.5.

Cyclohexane is air-oxidize at a temperature of 150 - 160°C and about 8 to 10 atm in the presence of a cobalt catalyst.

$$C_6H_{12} + 2O_2 \longrightarrow C_6H_{11}OH + C_6H_{11}O$$

The product is a cyclohexanol-cyclohexanone (ketone-alcohol, or KA) mixture. The mixture is distilled to recover unconverted cyclohexane which is recycled to the reactor feed. The resultant KA mixture may then be distilled for improved quality before being sent to the nitric acid oxidation stage. This process yields 75 to 80 mole percent KA, with a ketone to alcohol ratio of 1:2. The second step is the nitric acid oxidation of the cyclohexanol/cyclohexanone (KA) mixture. The reaction proceeds as follows:

$$C_6H_{11}OH + C_6H_{11}O + zHNO_3 \longrightarrow HOOC(CH_2)_4COOH + xN_2O + yNO$$

50 to 60% nitric acid in the presence of a copper-vanadium catalyst is reacted with the KA mixture in a reactor at 60 to 80 °C and 0.1 to 0.4 MPa. Conversion yields of 92 to 96 percent are attainable when using high-purity KA feedstock. As the reaction is highly exothermic, the heat of reaction is usually dissipated by maintaining a high ratio (40:1) nitric acid to KA mixture.

Upon reaction, nitric acid is reduced to nitrogen oxides: $NO_2$, NO,

$N_2O$, and $N_2$. The dissolved oxides are stripped from the reaction product using air in a bleaching column and subsequently recovered as nitric acid in an absorption tower. The $N_2$ and $N_2O$ are released to the atmosphere. Nitrogen oxides, entering the lower portion of the absorber, flow countercurrent to a water stream, which enters near the top of the absorber. Unabsorbed NO is vented from the top while diluted nitric acid is withdrawn from the bottom of the absorber and recycled to the adipic acid process.

The stripped adipic acid/nitric acid solution is chilled and sent to a crystallizer, where crystals of adipic acid are formed. The crystals are separated from the mother liquor using a centrifuge and transported to the adipic acid drying and/or melting facilities. The mother liquor is separated from the remaining uncrystallizedadipic acid in the product still and recycled to the reactors.

## 5.3. The Polymer Industry

## 5.3.1. Introduction

Industrial use of polymers started when Goodyear discovered the vulcanization of rubber in 1839. Polymer research rapidly spread throughout the world after 1930 and this led to the development of many synthetic polymers including nylon, polyethylene and polyvinyl chloride.

Polymers are high molecular weight compounds built up by the repetition of small chemical units known as monomers. They are either natural or synthetic. The natural polymers include rubber, cellulose, wool, starch and proteins.

The term polymer comes from two Greek words: "polys" which means "many" and "meros" which means "parts." A polymer is therefore a high molecular weight compound made up of hundreds or thousands of many identical small basic units (monomers) of carbon, hydrogen, oxygen or silicon atoms. The monomers are linked together covalently in a chemical process known as polymerization. This is illustrated in Fig 5.5.

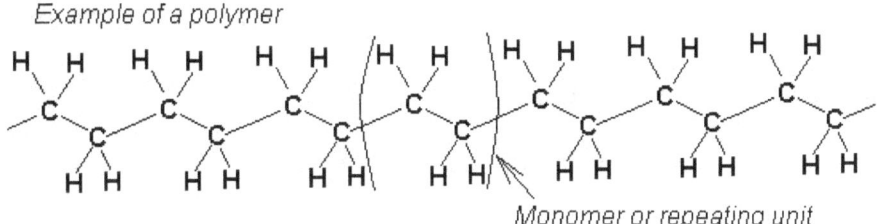

**Fig 5.5.**     A polymer chain

## 5.3.2. Classification of polymers

Polymers can be classified into three types:

1.  Linear polymers in which the repeating units are similar to the links in a very long chain They are known as polymer chains. An example is polyethylene

2.  Branched polymers in which some of the molecules are attached as side chains to the linear chains. The individual molecules are still discrete.

Three-dimensional cross-linked polymers in which branched chains are joined together by cross-linking in a process known as "curing". Vulcanization of rubber is a curing process.

## 5.3.3.   Polymer properties

Polymers have three main properties:

<u>Molecular weight</u>: The molecular weight of polymers is not fixed because of varying chain length.

<u>Crystallinity</u>: Because of high molecular weight and varying chain length, most po-lymers are amorphous and only semi-crystalline. Those with high crystallinity are tougher, more opaque, more resistant to solvents, higher density and sharply defined melting point.

<u>Glass transition temperature</u>: At low temperature, even amorphous polymers are hard and brittle (glass-like). As temperature is increased, kinetic energy increases. Howe-ver, motion is restricted to short-range vibrations and rotations as long as glass-like structure is retained. At a certain temperature called the *glass transition temperature,* a polymer loses glass-like properties. It becomes softer and more elastomeric but it does not melt. If heating is continued further, the polymer will lose elastomeric properties and melt to a flowable liquid.

### 5.3.4. Types of polymer products

*Plastics*

A plastic is a material that contains as an essential ingredient, an organic substance of a large molecular weight, is solid in its finished state, and, at some stage in its manufacture or in its processing into finished articles, can be shaped by flow.

In practice, a plastic is usually considered to be an amorphous or crystalline polymer which is hard and brittle at ordinary temperatures. If crystalline regions are present, they are randomly oriented.

<u>*Thermoplastics*</u>

A thermoplastic material is one which can be softened and molded on heating. They are elastic and flexible above a certain glass transition temperature. Nylon is a thermoplastic and it was the first commercial polymer to be made as a substitute for silk for making parachutes, vehicle tires, garments and many other products. Current uses include: fabrics, footwear, fishnets and carpets to mention but a few. The two special grades of Nylon are Nylon 6-6 and Nylon 6. Other thermoplastic materials and their uses are given in Table 5.2

**Table 5.2. Uses of various thermoplastic materials**

| Plastic type | Uses |
| --- | --- |
| Low density polyethylene (LDPE | Packaging films, wire and cable insulation, toys, flexible bottles, hou-Seware |
| High density polyethylene (HDPE) | Bottles, drums, pipes, films, sheets, wire and cable insulation |
| Polypropylene PP | Automobile and appliances parts, furniture, cordage, carpets, film Packaging |
| Polyvinyl chloride PVC | Construction, rigid pipes, flooring, wire and cable insulation, film and Sheet |
| Polystyrene | Packaging (foam and film), foam, insulation, insulation, appliances, Houseware |

## Thermosetting plastics

A thermosetting material is one which involves considerable crosslinking, so that the finished plastic cannot be made to flow or melt. Thermosetting plastics (thermosets) are polymer materials that cure or are made strong by addition of elements (e.g. sulphur) or addition of energy in form of heat (normally above 200°C) through some chemical reaction. Before curing process, they are usually in liquid form, powder or malleable forms that can be moulded to a final form or used as adhesives. The curing process transforms these materials into plastic or rubber through a cross linking process. The cross links produce a three dimensional rigid structure of the material with large molecular weight and a high melting point. The three dimensional network of bonds in thermosets generally makes them much stronger than thermoplastics. This makes them suitable for high temperature applications up to the decomposition tem-perature of the material. A thermoset material cannot be melted and reshaped after forming and curing and therefore cannot be recycled unlike thermoplastics. Examples of thermosets include: polyester resin, vulcanized rubber, bakelite and epoxy resin. Table 5.3. gives uses of various thermosetting plastics

**Table 5.3. Principal thermosetting plastics**

| Thermosetting plastic | Uses |
| --- | --- |
| Phenol-formaldehyde (PF) | Electrical and electronic equipment, automobile parts, utensils, handles, plywood adhesives, particle board Binder |
| Urea-formaldehyde (UF) | Similar to PF, textile treatment, coating |
| Unsaturated polyester (UP) | Construction, automobile parts, marine Accessories |
| Epoxy | Protective coating, adhesives, electri-cal and electronics, industrial flooring, material composites |
| Melamine-formaldehyde (MF) | Similar to UF, decorative panels, coun-ter and table tops, dinnerware |

In the fabrication of plastic objects, additives such as colourants, fillers, plasticizers, lubricants and stabilizers are commonly added to modify the physical and mechanical properties of the material.

**Elastomers**

An elastomer (or rubber) is a word having its origin from two words: "elastic" which means the ability to return to original shape when a force or stress is removed and "mero" which means "parts" implying many parts or monomers. Therefore, an essential requirement of an elastomer is that it must be elastic i.e. it must stretch rapidly under tension to several times its original length with little loss of energy as heat. Industrial elastomers have high tensile strength and high modulus of elasticity. They are amorphous polymers with considerable cross-linkages. The covalent cross-linka-ges make the elastomer to return to its original structure or shape when the stress is removed. Without cross-linkages or with short chains, the applied force would result in a permanent deformation.

They are usually thermosets that require vulcanization, but there are some which are thermoplastic. Elastomers include:

*   Nitrile rubber

*   Butyl rubber

*   Silicone rubber

*   Polyurethane rubber

*   Polysulphide rubber

*   Poly butadiene

*   Styrene-butadiene

*   Polyisoprene

*   Tetrafluoroethylene

*   Tetrafluoropropylene

*Adhesives*

The heated glue-pot which traditionally contained glues based on animal products such as hoof, horn and fish residues has been replaced by adhesives based on synthetic polymers. There is now a wide range of adhesives and sealants suited to a variety of tasks from polyvinyl acetate (PVA) wood, board and paper glues to two-part epoxide resins for rivet-less bonding of metal panels.

## Fibres

Animal fibres, such as wool or silk, and vegetable fibres, such as cotton, continue to be used although there is a wealth of synthetic fibres such as cellulose acetate and nylon, acrylic and polyester. Carbon fibres for making advanced composites are produced by heat treatment of polyacrylonitrile and other synthetic fibres.

## Films

Animal membranes were the only non-metallic film forming materials used before the availability of rubber and these found little application. The successful development of a drum for casting films from viscose led in the 1920s to the production of 'Cellophane'- still a widely used material. In the 1930s, unsupported PVC films were manufactured but it was not until polyethylene was available in the 1940s that the production of films for bagging materials became commonplace.

## Surface finishes

The paint industry was traditionally based on naturally occurring 'drying' oils such as linseed but since the 1930s these have gradually been replaced by synthetic polymers. Because of toxicity problems from using paints based on solvents, many more finishes are now water-based polymer emulsions.

### 5.3.5. Polyethylene

### 5.3.5.1. Introduction

There are three major classes of polyethylene. These are low density polyethylene (LDPE), high density polyethylene (HDPE) and linear low density polyethylene

(LLDPE). Pellets of these plastics are extruded and blown to produce film. This film is used for packaging and making plastic bags.

Ethylene is derived from either modifying natural gas (a methane, ethane, propane mix) or from the catalytic cracking of crude oil. In a highly purified form, it is piped directly from the refinery to a separate polymerization plant. Here, under the right conditions of temperature, pressure and catalysis, the double bond of the ethylene monomer opens up and many monomers link up to form polyethylene. In commercial polyethylene, the number of monomer repeat units ranges from 1000 to 10,000. Molecular weight ranges from 28,000 to 28,0000.

### 5.3.5.2. The Polyethylene Manufacturing Process

Today, polyethylene manufacturing processes are usually categorized into "high pressure" and "low pressure" operations. The former is generally recognized as producing conventional low density polyethylene (LDPE) while the latter makes high density (HDPE) and linear low density (LLDPE) polyethylenes. The difference between these polyethylene processes and types is outlined below.

## High pressure

Polyethylene was first produced by the high pressure process by ICI, Britain, in the1930's. They discovered that ethylene gas could be converted into a white solid by heating it at very high pressures in the presence of minute quantities of oxygen:

**Ethylene + 10 ppm oxygen** $\xrightarrow{1000 - 3000 \text{ bar}}$ **polyethylene**

**80 - 300 $^{o}$C**

The polymerization reaction which occurs is a random one, producing a wide distribution of molecule sizes. By controlling the reaction conditions, it is possible to select the average molecule size (or molecule weight) and the distribution of sizes around this average (molecular weight distribution). The chains are highly branched (at intervals of 20–50 carbons).

ICI named their new plastic "polythene" and found that they were able to produce it in a density range of about 0.915 to 0.930g cm$^{3}$. It is known today as LDPE and has its single biggest usage in blown film.

## Low pressure

The initial discovery of LDPE was an accident. So was the discovery of HDPE in 1952. Researchers in Germany and Italy had succeeded in making a new aluminium based catalyst which permitted the polymerization of ethylene at much lower pressures than the ICI process:

Ethylene $\xrightarrow{10 - 80 \text{ bar}}$ polyethylene

**70 - 300 $^{o}$C, Al catalyst**

The product from this process was found to be much stiffer than previous products and had a density range of about 0.940 - 0.970g cm$^3$. The increased stiffness and density were found to be due to a much lower level of chain branching. The new HDPE was found to be composed of very straight chains of ethylene with a much narrower distribution of molecular weights (or chain lengths) and a potentially very high average chain length.

In the late 1950's, DuPont Canada first applied the low pressure process to the production of LLDPE. LLDPE is made by copolymerizing of ethylene with a small amount of another monomer, typically butene, hexene or octene.

The most common method used in industry is to polymerize ethylene by means of a fluidized reactor bed. A fluidized reactor bed consists of metallic catalyst particles that are 'fluidized' by the flow of ethylene gas. This means that the catalyst particles are suspended in the ethylene fluid as ethylene gas is pumped from the bottom of the reactor bed to the top.

Before the late 1970's an organic peroxide catalyst was employed to initiate polymerization. However, because the organic peroxide catalyst is not as active as the metallic catalyst, pressures in excess of 100 times the pressure required with metallic catalysts were necessary.

Before ethylene is sent to the fluidized bed, it must first be compressed and heated.

Pressures in the range of 100-300 pounds per square inch (psi) and a temperature of 100°C are necessary for the reaction to proceed at a reasonable rate. The catalyst is also pumped with the ethylene stream into the reactor. This is because polyethylene molecules remain stuck to the catalyst particle from which they were produced thus incorporating the catalyst within the polyethylene product. Hence the need to replenish the "consumed " catalyst

The conversion of ethylene is low for a single pass through the reactor and it is necessary to recycle the unreacted ethylene. Unreacted ethylene gas is removed off the top of the reactor. After purification, ethylene gas is then recompressed and recycled back into the reactor. Granular polyethylene is gradually removed from the bottom of the reactor as soon as reasonable conversions have been achieved. Typically, a residence time of 3 to 5 hours results in a 97% conversion of ethylene.

Polyethylene comes out of the reactor as granular powder, which is then melted and flows through a film extruder.

Whatever the type of polyethylene produced, the end product is usually available in the form of small pellets, varying in shape (spherical, rectangular, cylindrical) depending upon the manufacturer's equipment. During the manufacture of polythene products, it is melted to flow through a film extruder.

LDPE is the preferred packaging material due to its limp feel, transparency, toughness, and the ability to rapidly take up the shape of the contents of the bag. The garbage bag is just one of many widely practical uses of plastic bags.

Polyethylene film, produced by blown film extrusion, is commonly used for packaging of foodstuffs and other products. The thickness of the film produced tends to be from 20 - 200 μm.

### 5.3.6.    Styrene Butadiene Rubber (SBR)

### 5.3.6.1. Introduction

Emulsion polymerized styrene-butadiene rubber (E-SBR) is one of the most widely used polymers in the world today. Emulsion SBR is employed in many demanding applications, which enhance the quality of life and contribute significantly to our economy and standards of living.

In the 1930's, the first emulsion polymerized SBR known as Buna S was prepared by I. G. Farbenindustrie in Germany. The U. S. Government in 1940 established the Rubber Reserve Company to start a stockpile of natural rubber and a synthetic rubber

program. These programs were expanded when the United States entered World War II. The synthetic rubber efforts were initially focused on a hot polymerized (41°C) E-SBR.

Production of a 23.5% styrene and 76.5% butadiene copolymer began in 1942. Cold polymerized E-SBR (5°C), that has significantly better physical properties than hot polymerized SBR, was developed in 1947.

### 5.3.6.2. Uses

SBR is widely used for rubber belting, hose, flooring, molded goods, rubber soles, coated fabrics etc. It is compatible with natural rubber and has equal performance for automobile tyres. But it is inferior to natural rubber for heavy duty truck tyres.

### 5.3.6.3. Manufacturing Process

SBR is produced by the copolymerization of butadiene with styrene in the approxi-mate proportion of 3:1 by weight.

In the emulsion process, which produces general purpose grades, the feedstocks are suspended in a large proportion of water in the presence of an initiator or a catalyst and a stabiliser. A continuous process is employed.

In the solution process, the copolymerisation proceeds in a hydrocarbon solution in the presence of an organometallic complex. This can be either a continuous or batch process.

The emulsion polymerization process has several advantages. It is normally used under mild reaction conditions that are tolerant to water and requires only the absence of oxygen.

The process is relatively robust to impurities and amenable to using a range of monomers. Additional benefits include the fact that emulsion polymerization gives high solids contents with low reaction viscosity and is a cost-effective process. The physical state of the emulsion (colloidal) system makes it easy to control the process. Thermal and viscosity problems are much less significant than in bulk polymerization.

Table 5.4 shows the raw materials required in the polymerization of E-SBR, They include the monomers styrene and butadiene, water, emulsifier, initiator system, modifier, shortstop and a stabilizer system. The original polymerization reactions were charged out in batch reactors in which all the ingredients were loaded to the reactor and the reaction was shortstopped after it had reached the desired conversion. Current commercial productions are run continuously by feeding reactants and polymerizing through a chain of reactors before shortstopping at the desired monomers conversion.

The monomers are continuously metered into the reactor chains and emulsified with the emulsifiers and catalyst agents.

In cold polymerization, the most widely used initiator system is the redox reaction between chelated iron and organic peroxide using sodium formaldehyde sulfoxide (SFS) as reducing agent as shown in the following reactions.

$$Fe(II)EDTA + ROOH \longrightarrow Fe(III)EDTA + RO + OH.$$

$$Fe(III)EDTA + SFS \longrightarrow Fe(II)EDTA$$

In hot polymerization, potassium peroxydisulfate is used as an initiator.

Mercaptan is added to furnish free radicals and to control the molecular weight distribution by terminating existing growing chains while initiating a new chain. The thiol group acts as a chain transfer agent to prevent the molecular weight from attaining the excessively high values possible in emulsion systems. The sulfur-hydrogen bond in the thiol group is extremely susceptible to attack by the growing polymer radical and thus loses a hydrogen atom by reacting with polymer radicals as shown below. The RS formed will continue to initiate the growth of a new chain. The thiol prevents gel formation and improves the processability of the rubber.

$$P\cdot + RSH \longrightarrow P\text{-}H + RS\cdot$$

$$RS\cdot + M \longrightarrow RS\text{-}M\cdot$$

## Table 5.4.  Typical Recipe for SBR Emulsion Polymerization

| Component | Parts by Weight | |
|---|---|---|
| | Cold | HOT |
| Styrene | 25 | 25 |
| Butadiene | 75 | 70 |
| Water | 180 | 180 |
| Emulsifier (FA,RA, MA) | 5 | 5 |
| Dodecyl mercaptan | 0.2 | 0.8 |
| Cumenehydroperoxide | 0.17 | - |
| $FeSO_4$ | 0.017 | - |
| EDTA | 0.06 | - |
| $Na4P_2O_7.10H_2O$ | 1.5 | - |
| Potassuimpersulfate | | 0.3 |
| SFS | | 0.1 |
| Stabilizer | | Varies |

During polymerization, parameters such as temperature, flow rate and agitation are controlled to get the right conversion. Polymerization is normally allowed to proceed to about 60% conversion in cold polymerization and 70% in hot polymerization before it is stopped with a shortstop agent that reacts rapidly with the free radicals. Some of the common shortstopping agents are sodium dimethyldithiocarbamate and diethyl hydroxylamine.

Once the latex is properly shortstopped, the unreacted monomers are stripped off the latex. Butadiene is stripped by degassing the latex by means of flash distillation and reduction of system pressure. Syrene is removed by steam stripping the latex in a column. The latex is then stabilized with the appropriate antioxidant and transferred to blend tanks. In the case of oil-extended polymers or carbon black masterbatches, these materials are added as dispersions to the stripped latex. The latex is then transferred to finishing lines to be coagulated with sulfuric acid, sulfuric acid/sodium chloride, glue/sulfuric acid, aluminumsulfate, or amine coagulation aid. The type of coagulation system is selected depending on the end-use of the product. Sulfuric acid/sodium chloride is used for general purpose. Glue/sulfuric acid is used for electrical grade and low water sensitivity SBR. Sulfuric acid is used for coagulations where low-ash-polymer is required. Amine coagulating aids are used to improve coagulation efficiency and reduce production plant pollution. The coagulated crumb is then washed, dewatered, dried, baled and packaged.

**Unit objectives**

At the end of this unit you should be able to:

a. Discuss the occurrence and extraction of petroleum

b. Explain the purposes and application of fractional distillation, catalytic cracking and catalytic reforming during petroleum processing

c. Describe using equations and flow diagrams, the manufacture of some petrochemicals, namely, phthalic anhydride and adipic acid

d. Categorize polymerization reactions, polymers and polymer products

e. Describe the uses of various plastics

f.  Explain how polyethylene and styrene butadiene rubber are manufactured

## Summary of Chapter

This is the first of two units dealing with organic chemical industries. We start with petroleum refining because petroleum and its primary derivatives are the backbone of modern organic chemical industries. We shall then study the manufacture of selected petrochemicals namely phthalicanydride and adipic acid. Next, we shall look at various types of polymers, their uses and how some of them are manufactured.

## References

1.  George T. A. (1977). *Shreve's Chemical Process Industries. 5th edn.* McGRAW-HILL INTERNATIONAL EDITIONS. Chemical Engineering Series. Singapore.

2.  Stephenson R.M. (1966). Introduction to the Chemical Process Industries, Reinhold Publishing Corporation, New York

3.  Groggins P.H. (1958). Unit Processes in Organic Synthesis, 5th Edition, McGraw-Hill Book Company, New Delhi.

4.  Gerhartz, W. (Editor), (1987). Ullmann'sEncyclopaedia of Industrial Chemistry, 5th Edition, VCH VerlagsgesellschaftmbH, Weinheim.

## List of relevant resources

*   Computer with internet facility to access links and relevant copywrite free resources

*   CD-Rom accompanying this module for compulsory reading and

demonstrations

- Multimedia resources like video, VCD and CD players

## List of relevant useful links

http://www.setlaboratories.com

http:/www.lloyadministerheavyoil.co

http://chem.ucalgary.ca/course/351/carey/Ch08/ch-

4.html http://chemguide.co.uk

http://pslc.ws/

http://www.wisegeek.com

http://faculty.washington.edu/finlayso/Polyeth/Group_B/them

er.html

From these sites, you will find a wealth of information on the technology of petroleum refining with associated process flow diagrams. The chemistry including reaction mechanisms of most of the processes are given. You will also find a wide variety of chemical products and their production methods.

# GLOSSARY

**Alkylation** is the introduction of an alkyl radical by substitution or addition into anorganic compound.

**Antibiotics** are chemical substances that can inhibit the growth of, and even destroy,harmful microorganisms.

**Catalytic cracking** is the breaking up of complex hydrocarbons into simpler molecules in order to increase the quality and quantity of lighter, more desirable products and decrease the amount of residuals.

**Catalytic reforming** is a process used to convert low-octane naphthas into high-octane compounds such as toluene, benzene, xylene, and other aromatics which are useful in gasoline blending and petrochemical processing.

**Emulsion polymerization** is a free radical polymerization that take place in anemulsion consisting of water, monomer, surfactant and other additives.

**Fermentation** is a reaction wherein a raw material is converted into a product by theaction of micro-organisms or by means of enzymes.

**Fertilizers** are chemical compounds given to plants to promote growth

**Industrial chemistry** as the branch of chemistry which applies physical and chemical procedures towards the transformation of natural raw materials and their derivatives to products that are of benefit to humanity.

**Material balance is the application of the law of conservation of mass in the form of equations** to satisfy balances of total masses, components and atomicspecies through a process.

**Ore dressing** is the pretreatment of mineral ores by mainly physical processes toeffect the concentration of valuable minerals and to render the enriched material into the most suitable physical condition for subsequent operations.

**Plastic** is a material that contains as an essential ingredient, an organic substance ofa large molecular weight, is solid in its finished state, and, at some stage in its manufacture or in its processing into finished articles, can be shaped by flow.

**Surfactant** is a compound consisting of a long, linear, non-polar (hydrophobic) 'tail'with a polar (hydrophilic) 'head' which lowers the surface tension of water and allows oil to form an emulsion with water

**Unit operations** are the physical treatment steps employed in chemical processes totransform raw materials and products into required forms.

**Unit processes** are the chemical transformations or conversions that are performedin a process.